# NEED to KNOW

## AQA A-LEVEL GEOGRAPHY

Key facts at your fingertips

David Redfern

## HODDER EDUCATION
AN HACHETTE UK COMPANY

Hachette UK's policy is to use papers that are natural, renewable and recyclable products and made from wood grown in sustainable forests. The logging and manufacturing processes are expected to conform to the environmental regulations of the country of origin.

Orders: please contact Bookpoint Ltd, 130 Park Drive, Milton Park, Abingdon, Oxon OX14 4SE. Telephone: (44) 01235 827827. Fax: (44) 01235 400401. Email: education@bookpoint.co.uk

Lines are open from 9 a.m. to 5 p.m., Monday to Saturday, with a 24-hour message answering service. You can also order through our website: www.hoddereducation.co.uk

ISBN: 978 1 5104 2848 5

© David Redfern

First published in 2018 by
Hodder Education,
An Hachette UK Company
Carmelite House
50 Victoria Embankment
London EC4Y 0DZ

Impression number     10 9 8 7 6 5 4 3 2 1

Year     2022 2021 2020 2019 2018

Typeset in India by Aptara Inc.

Printed in Spain

A catalogue record for this title is available from the British Library.

MIX
Paper from
responsible sources
FSC™ C104740

# Contents

# Getting the most from this book

This *Need to Know* guide is designed to help you throughout your course as a companion to your learning and a revision aid in the months or weeks leading up to the final exams.

The following features in each section will help you get the most from the book.

## You need to know

Each topic begins with a list summarising what you 'need to know' in this topic for the exam.

## Exam tip

Key knowledge you need to demonstrate in the exam, tips on exam technique, common misconceptions to avoid and important things to remember.

## Key terms

Definitions of **highlighted** terms in the text to make sure you know the essential terminology for your subject.

## Do you know?

Questions at the end of each topic to test you on some of its key points. Check your answers here: www.hoddereducation.co.uk/needtoknow/answers

## End of section questions

Questions at the end of each main section of the book to test your knowledge of the specification area covered. Check your answers here: www.hoddereducation.co.uk/needtoknow/answers

# ① Water and carbon cycles

## 1.1 Water and carbon cycles as natural systems

**You need to know**
- what constitutes a system
- how systems operate with inputs, outputs, stores and flows
- the concept of dynamic equilibrium
- the various forms of feedback mechanisms

## Systems: inputs, outputs, stores and flows

Systems occur in both physical geography and human geography. They have several common features, as shown in Table 1.

**Table 1 Features of systems**

| Inputs | Outputs | Stores | Transfers or flows |
|---|---|---|---|
| Elements that enter a system to be processed | Outcome(s) of processing within the system | Amounts of energy or matter held, and not transferred until the appropriate processes are in place to move them | Movements of energy or matter through the system, which enable inputs to become outputs |

## Dynamic equilibrium

When there is a balance between the inputs and the outputs, the system is said to be in a state of dynamic equilibrium.

## Feedback mechanisms

Feedback occurs when one of the elements of the system changes, such as an input. The state of the store changes and the equilibrium is upset. There are two types of feedback:
- positive feedback, where a change causes a further (or snowball) effect, continuing or even accelerating the original change
- negative feedback, which acts to lessen the effect of the original change and ultimately to reverse it

### Key terms

**System** A set of interrelated components that are connected together to form a working unit or unified whole.

**Dynamic equilibrium** The balanced state of a system when opposing forces, or inputs and outputs, are equal.

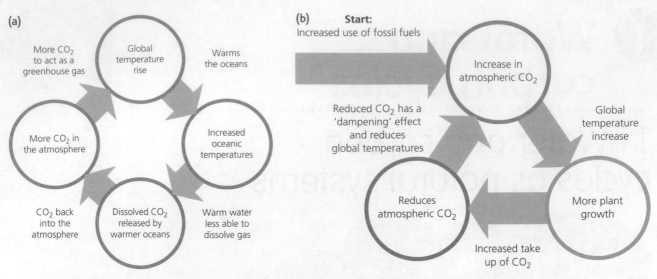

**Figure 1 Examples of feedback in natural systems: (a) positive feedback, (b) negative feedback**

## Do you know?

1 Outline the differences between 'inputs' and 'outputs' in a system.
2 Explain the difference between 'negative' and 'positive' feedback mechanisms.

## Exam tip

Make sure you can use systems theory, including feedback, in a number of contexts across geography.

# 1.2 The water cycle

## You need to know

- the global distribution and size of major water stores
- the processes driving change in the water cycle
- how drainage basins act as open systems
- the key elements of the flood (storm) hydrograph
- the factors affecting change in the water cycle over time

# Major water stores
## Global distribution

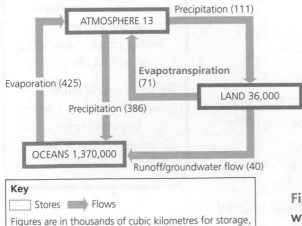

**Key**
☐ Stores ➡ Flows
Figures are in thousands of cubic kilometres for storage, and thousands of cubic kilometres/year for flows.

**Figure 2 The global water cycle: stores and annual flows**

## Key term

**Evapotranspiration** The combined water gain by the atmosphere through evaporation (90%) and transpiration (10%).

Key features:

- 97% of the world's water is saline sea water
- almost 80% of total fresh water is locked up in ice and glaciers
- another 20% of fresh water is in the ground
- surface fresh water sources, such as rivers and lakes, constitute only about 1/150th of 1% of total water

## The major water stores

The hydrosphere:

- oceans hold the vast majority of all water on Earth
- oceans supply about 90% of the evaporated water that goes into the water cycle
- the amount of water in the oceans has changed:
  - □ during the last ice age sea levels were lower, by as much as 120 m, as more ice caps and glaciers formed
  - □ during the last major global 'warm spell', 125,000 years ago, the seas were about 5.5 m higher than they are now

The atmosphere:

- contains a very small store of water (0.001% of the Earth's total water)
- is the main vector that moves water around the globe (Figure 3)
- clouds are a visible manifestation of atmospheric water, but clear air also contains water vapour

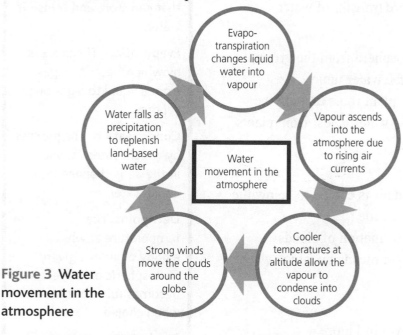

**Figure 3** Water movement in the atmosphere

The cryosphere:

- Antarctica holds almost 90% of the Earth's ice mass
- the Greenland ice cap contains 10% of the total global ice mass
- collectively, ice caps and glaciers cover about 10% of the Earth's surface

> ### Exam tip
>
> It is important that proportions, or percentages, of relative amounts of water are learnt and understood in this section.

- ice shelves in Antarctica cover over 1.6 million km² (an area the size of Greenland), covering 11% of its total area
- sea ice (frozen sea water) surrounds several polar regions of the world. On average sea ice covers up to 25 million km², an area 2.5 times the size of Canada

The lithosphere:

- surface fresh water (glaciers, rivers, streams, lakes, reservoirs and wetlands) represents about 2.5% of all water on Earth
- 20% of all accessible surface fresh water is in Lake Baikal (Russia)
- another 20% is stored in the Great Lakes of North America
- rivers hold only about 0.006% of total fresh water reserves
- large quantities of water are also held deep underground in aquifers
- water from aquifers can take thousands of years to move back to the surface; some never does

## Exam tip

Questions will make use of the terms *lithosphere*, *hydrosphere*, *cryosphere* and *atmosphere*. Make sure you do not get them confused.

# Processes driving change

Evaporation:

- heat (energy) is necessary for evaporation to occur — it is used to break the bonds that hold water molecules together
- this is why water easily evaporates at boiling point (100°C) and evaporates much more slowly at freezing point
- evaporation from the oceans is the primary mechanism supporting the surface-to-atmosphere transfer of water

Evapotranspiration:

- the combined water loss to the atmosphere from the ground surface and the capillary fringe of the water table
- the transpiration of groundwater by plant roots tapping the capillary fringe of the water table — water is lost from plants through the stomata in their leaves

Condensation and cloud formation:

- condensation occurs when saturated air is cooled to below the dew point, usually due to a rise in altitude
- condensation is responsible for the formation of clouds
- clouds may produce precipitation — a means by which water returns to the Earth's surface

Precipitation:

- one precipitation mechanism is shown in Figure 4
- another mechanism (the Bergeron–Findeisen process) leads to the rapid growth of ice crystals at the expense of the water vapour present in a cloud
- these crystals fall as snow, and melt to become rain

## Key terms

**Ice shelf** A floating extension of land ice.

**Aquifer** A permeable rock that can store and transmit water.

**Evaporation** The process by which water changes from a liquid to a gas or vapour.

**Condensation** The process by which water vapour in the air is changed into liquid water.

**Dew point** The temperature at which a body of air at a given atmospheric pressure becomes fully saturated when cooled.

**Precipitation** Water released from clouds in the form of rain, freezing rain, sleet, snow or hail.

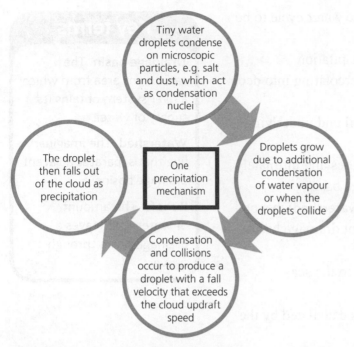

**Figure 4 One precipitation mechanism**

Snowmelt and runoff:

- runoff from snowmelt varies in importance both geographically and over time
- in areas with colder climates, much springtime flow in rivers is attributable to melting snow and ice

# Drainage basins

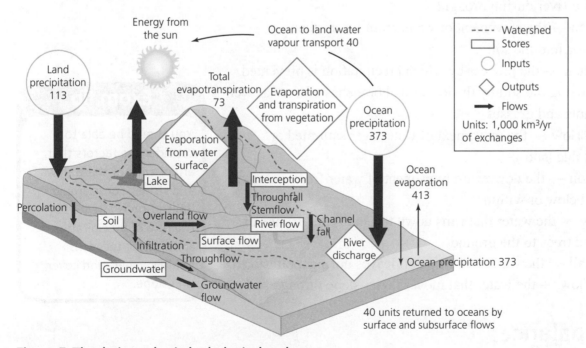

**Figure 5 The drainage basin hydrological cycle**

The **drainage basin** hydrological cycle allows the water cycle to be examined at a local scale:

- inputs include energy from the Sun and precipitation
- outputs include evapotranspiration, water percolating into deep groundwater stores, and runoff into the sea
- stores include vegetation, the ground, the soil and underlying bedrock
- transfers take place between any of these stores and ultimately into the channels of the rivers of the drainage basin
- drainage basins are bounded by high land (**watershed**), beyond which any precipitation falls into the adjacent drainage basin
- all flows lead water to the nearest river
- the river transfers water by its channel flow to the sea (**runoff**) — measured by its discharge
- for any river at a given location, discharge is calculated by the following:

     discharge ($Q$) = average velocity ($V$) × cross-sectional area ($A$)

  □ the unit is cumecs, measured in $m^3 s^{-1}$

## Stores and flows in a drainage basin

See Figure 5.

- groundwater store — water that collects underground in pore spaces in rock
- groundwater flow — the movement of groundwater. This is the slowest transfer of water within the drainage basin and provides water for a river during drought
- infiltration — the movement of water from the surface downwards into the soil
- interception — the process by which precipitation is prevented from reaching the soil by the leaves and branches of trees as well as by plants and grasses
- overland flow — the movement of water over saturated or impermeable land
- percolation — the downward movement of water from soil into the rock below or within
- stemflow — the water that runs down the stems and trunks of plants and trees to the ground
- throughfall — the water that drips off leaves during a rainstorm
- throughflow — the water that moves down-slope through soil

## Water balance

This is the relationship between inputs of precipitation ($P$) and outputs in the form of evapotranspiration ($E$) and runoff ($Q$)

> ### Key terms
>
> **Drainage basin** The catchment area from which a river system obtains its supply of water.
>
> **Watershed** The imaginary line that separates adjacent drainage basins.
>
> **Runoff** The amount of water that leaves a drainage basin through a river.

> ### Exam tip
>
> You should be able to consider the factors that affect each of these flows and stores. For example, infiltration is affected by the rate of precipitation, soil type, antecedent rainfall, vegetation cover and slope.

together with changes to the amounts of water held in storage within the soil and groundwater ($\Delta S$):

$$P = E + Q + \Delta S$$

The water balance is shown through a soil water budget graph (Figure 6). This shows the balance between precipitation and **potential evapotranspiration**.

### Key terms

**Potential evapotranspiration** The amount of water that could be evaporated or transpired from an area assuming there is sufficient water available.

**Hydrograph** A graph of river discharge over time.

The soil moisture store is now all used up. Any precipitation is likely to go straight into the soil rather than travel to the river channel. River levels will fall. Some rivers dry up completely.

As it warms up potential evapotranspiration exceeds precipitation. The water store is being used by plants (**utilisation**).

There is a **deficit** of soil water. Plants either wilt or have adaptations to survive dry conditions.

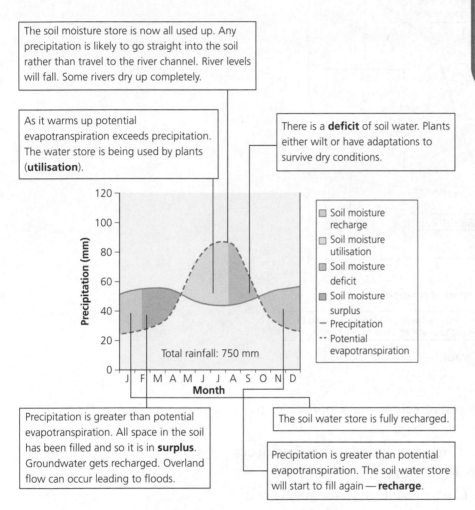

Precipitation is greater than potential evapotranspiration. All space in the soil has been filled and so it is in **surplus**. Groundwater gets recharged. Overland flow can occur leading to floods.

The soil water store is fully recharged.

Precipitation is greater than potential evapotranspiration. The soil water store will start to fill again — **recharge**.

**Figure 6 Soil water budget graph for eastern England**

# The flood (storm) hydrograph

A flood (storm) **hydrograph** (Figure 7):
- is a graph of the discharge of a river before, during and following a storm event, i.e. its runoff variation
- can help predict how a river might respond to a rainstorm, which in turn can help in managing a river
- is described as 'flashy' when water is transferred to a river quickly, meaning it responds rapidly to the storm and often leads to flooding

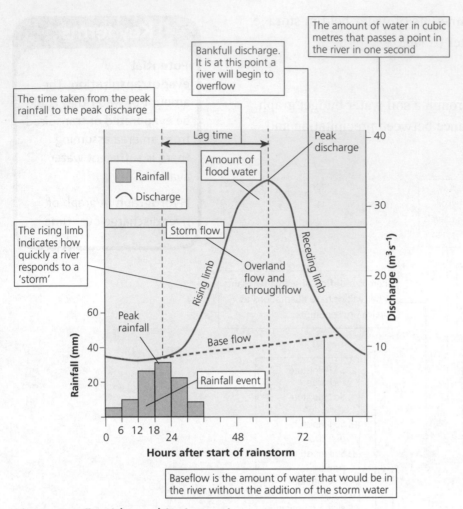

**Figure 7 A flood (storm) hydrograph**

A number of natural variables can affect the shape of the storm hydrograph:

■ antecedent rainfall — rain falling on saturated ground produces a steep rising limb and a shorter lag time

■ snowmelt — the release of large amounts of water greatly and rapidly increases discharge, especially if the surface is frozen as this reduces infiltration

■ vegetation — in summer, deciduous trees have more leaves so interception is higher, discharge lower and lag time longer

■ basin shape — water takes less time to reach the river in a circular drainage basin than in an elongated one

■ slope — steep-sided drainage basins allow water to get to a river more quickly than an area of gentle slopes

■ geology — permeable rocks allow percolation to occur, which slows down transfer of water; impermeable rocks reduce percolation, have greater amounts of overland flow and hence greater discharges and shorter lag times

## Exam tip

Consider how urban growth can affect a storm hydrograph.

# Changes in the water cycle

## Natural variations

Storm events resulting in a large input of water create:

- an extremely steep rising limb on the hydrograph, and a very short lag time
- an equally steep recession limb — the river returns to normal conditions in hours

Periods of cyclonic rainfall can add substantial amounts of water to a river catchment. They:

- have a timeline measured in hours/days from the onset of rainfall to the flood peak
- have a similar period for the recovery back to normal conditions
- can cause multiple flood events in the same catchment that can last several weeks

Seasonal floods occur where there is:

- seasonal rainfall (such as the Asian monsoon)
- seasonal snowmelt on a massive scale

## Human impacts

See Table 2.

**Table 2 Impacts of farming practices, land-use changes and water abstraction**

| Farming practices, e.g. arable farming | Land-use change, e.g. deforestation | Water abstraction |
|---|---|---|
| In late autumn, winter and early spring, crops are dormant and the soil is relatively bare | Removes water-absorbent forests, which trap and transpire rainfall, and replaces them with farmland | People make use of the water in aquifers: |
| Rain falling on these surfaces will not be intercepted by vegetation; overland flow rates are relatively high | Creates a significant increase in both the volume of water reaching a river and the speed with which it travels | <ul><li>in some places they pump water out faster than nature replenishes it</li><li>the water table can be lowered by excessive pumping</li><li>wells can 'go dry' and become useless</li></ul> |
| In late spring and summer, arable landscapes have fully established crops that can intercept greater proportions of the rainfall | However, some suggest the main impact is limited to the felling period: <ul><li>overland flow exists when tracks are driven through the forest and heavy machinery is used that compacts the soil</li><li>runoff patterns return to their prior state when land reverts to scrub or pasture</li></ul> | In places where the water table is close to the surface and where rocks are permeable, aquifers can be replenished artificially |
| This reduces peak flows, extending lag times | | |

## Do you know?

1 Explain the concept of dynamic equilibrium in relation to a drainage basin.

2 What are the differences between sea ice and an ice shelf?

3 Identify and explain the factors that determine transpiration rates.

4 Explain two ways in which overland flow is created.

5 Explain how variations in precipitation can impact the water cycle.

## Exam tip

You are required to study a river catchment at a local scale to illustrate and analyse how the impact of precipitation on stores and transfers has implications for sustainable water supply and/or flooding. This can be a classroom-based activity or based on fieldwork.

# 1.3 The carbon cycle

## You need to know

- the distribution and size of major stores of carbon
- the factors and processes driving change in the magnitude of these stores over time and space
- changes in the carbon cycle over time due to natural variations and human impacts
- the carbon budget and its impact on global climate

## Stores of carbon

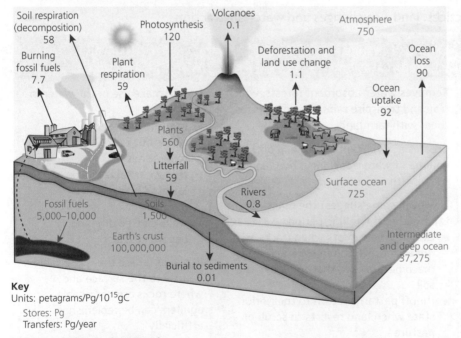

**Key**
Units: petagrams/Pg/$10^{15}$gC
  Stores: Pg
  Transfers: Pg/year

**Figure 8 The carbon cycle**

## Exam tip

Be aware of the relative sizes and rates of the various carbon cycle processes (Figure 8). It may be difficult to remember the numbers involved, but an understanding of relative scale is important.

See Figure 8.

- most of Earth's carbon, over 100 million petagrams (PgC), is stored in the lithosphere
- much of this is in fossil fuel rocks (coal and oil), and limestones

- the remainder is in the hydrosphere (38,000 PgC), the atmosphere (750 PgC), the biosphere (3000 PgC), and the cryosphere (mostly within the permafrost, amount unknown)
- carbon flows between each of these stores in a complex set of exchanges
- any change that shifts carbon out of one store puts more carbon in another store(s)

# Changes in the magnitude of stores

## Weathering

Key features:
- atmospheric $CO_2$ combines with water vapour to form a weak carbonic acid that falls as rain
- this acid dissolves rocks (chemical **weathering**) and releases calcium, magnesium, potassium, and sodium ions
- plants, through their growth, also break up surface granites, and microorganisms hasten the weathering with enzymes and organic acids in the soil coupled with the carbonic acid

## Carbon sequestration in oceans and sediments

See Figure 9.

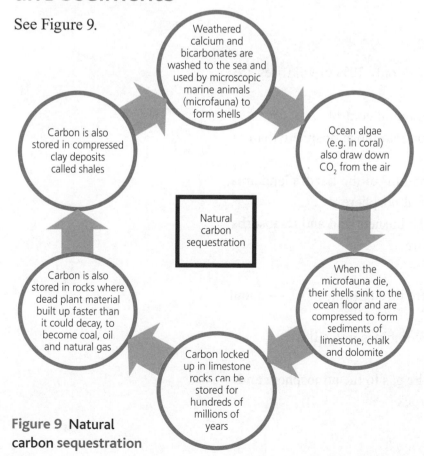

**Figure 9 Natural carbon sequestration**

## Photosynthesis, respiration, decomposition and combustion

Key points:

- plants and phytoplankton are key components of the carbon cycle
- during photosynthesis, plants absorb $CO_2$ and sunlight to create glucose and other sugars for building plant structures
- phytoplankton also take $CO_2$ from the atmosphere by absorbing it into their cells
- using energy from the Sun, both plants and phytoplankton combine $CO_2$ and water to form carbohydrate ($CH_2O$) and oxygen

Carbon can be returned to the atmosphere by:

- respiration — plants break down the carbohydrate to get the energy they need to grow; animals (and people) eat the plants and break down the sugar to get energy
- decomposition — plants and plankton die, decay, and are eaten by bacteria at the end of the growing season
- combustion — natural fire consumes plants

For these three processes, the $CO_2$ released in the reaction usually ends up in the atmosphere.

> ### Key term
>
> **Phytoplankton**
> Microscopic organisms in oceans.

> ### Exam tip
>
> Make sure you can distinguish between each of the terms: *photosynthesis, respiration, decomposition* and *combustion*.

# Changes over time

## Natural variations

Wildfires:

- can be caused by lightning strikes — only 10% of wildfires are started this way
- most are started by humans but go out of control
- forest fires can release more carbon into the atmosphere than forests can capture
- every year wildfires burn 4 million km² of the Earth's land area, and release tonnes of $CO_2$ into the atmosphere
- new vegetation then moves onto the burned land and re-absorbs much of the $CO_2$ released by the fire

Volcanic activity:

- carbon is emitted to the atmosphere through volcanoes — about 0.1 Pg of $CO_2$ per year
- during subduction, heated rock recombines into silicate minerals, releasing $CO_2$
- when volcanoes erupt, they vent the gas to the atmosphere and cover the land with fresh silicate rock

# Human impacts

Hydrocarbon fuel extraction and burning:

- people have influenced the carbon cycle through mining and subsequently burning **fossil fuels**
- some fossil fuels are volatile materials with low carbon to hydrogen ratios, such as methane and liquid petroleum
- some fossil fuels are non-volatile materials composed of almost pure carbon, such as anthracite coal
- burning fossil fuels produces around 21 Pg of $CO_2$ per year but natural processes can absorb only a proportion of this, resulting in a net increase of 8.5 Pg of atmospheric $CO_2$ per year
- $CO_2$ is a **GHG** that enhances atmospheric heating and contributes to climate change — a major environmental concern

Farming practices — examples of impact:

- tillage of cropland soils can cause them to lose carbon — it increases aeration and soil temperatures, making soil aggregates more susceptible to breakdown and making organic material more available for decomposition
- crop rotation, residue management, reduction of soil erosion and improvement of irrigation can increase carbon levels in soils
- rice cultivation and livestock farming are two sources of methane — alteration of rice cultivation practices and changes to livestock feed are therefore potential management practices that could reduce methane sources

Deforestation:

- see Figure 10

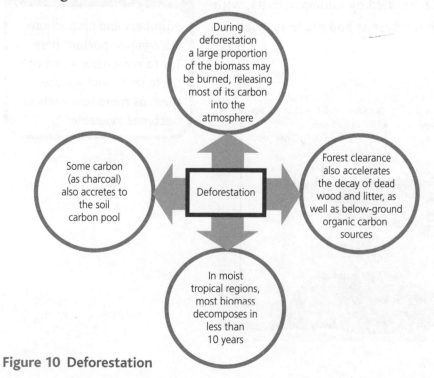

**Figure 10 Deforestation**

Land-use changes:
- The IPCC estimates that all land-use changes, including deforestation, contribute a net 1.6 PgC per year to the atmosphere
- other examples of land-use changes include:
  - ☐ drainage of wetlands — soils become exposed to oxygen, and carbon stocks (which are resistant to decay under the anaerobic conditions prevalent in wetland soils) can be lost by aerobic respiration
  - ☐ afforestation — tree plantations increase carbon sequestration

**Exam tip**

You are advised to be aware of what is being done globally to reduce the impact of deforestation and land-use change. Research the work of the UN-REDD+ scheme.

# The carbon budget

Key points:
- $CO_2$ is the single most important anthropogenic GHG in the atmosphere, contributing 65% of **radiative forcing**
- the current **carbon budget** shows a net gain of 4.4 PgC per year in the atmosphere (Figure 11)
- rising levels of $CO_2$ and other GHGs in post-industrial times are fuelling fears of climate change through atmospheric warming
- atmospheric $CO_2$ reached 142% of the pre-industrial level in 2013, mainly due to emissions from fossil-fuel combustion and cement production
- relatively small contributions to increased $CO_2$ come from deforestation and other land-use change, although the net effect of terrestrial biosphere fluxes is as a sink — 2 PgC per year
- the average increase in atmospheric $CO_2$ from 2003 to 2013 corresponded to 45% of the $CO_2$ emitted by human activity, with the remaining 55% removed by the oceans and the terrestrial biosphere

**Key terms**

**IPCC** The Intergovernmental Panel on Climate Change.

**Radiative forcing** The difference between incoming solar radiation (insolation) absorbed by the Earth and the energy radiated back out into space.

**Carbon budget** The balance of exchanges between the four major stores of carbon.

**Exam tip**

Numbers and proportions are again important here. Try to remember some of these facts, and update them as more information becomes available.

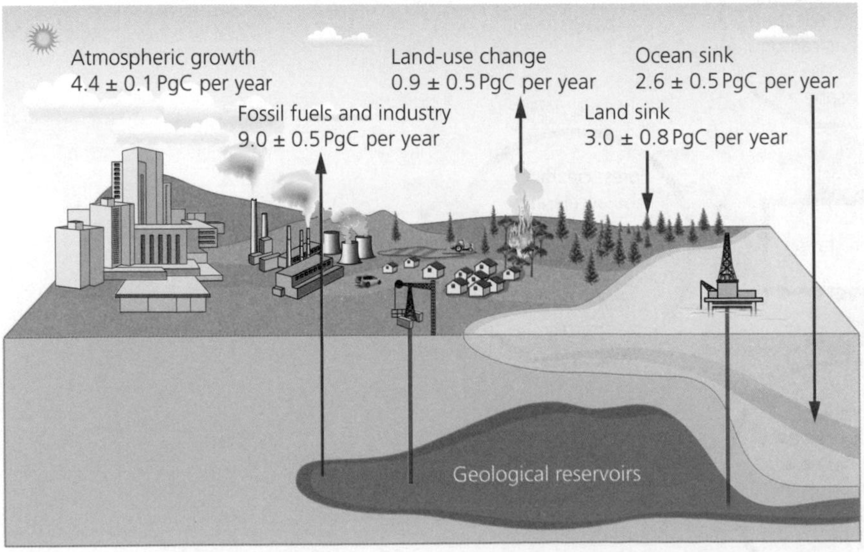

Atmospheric growth
4.4 ± 0.1 PgC per year

Land-use change
0.9 ± 0.5 PgC per year

Ocean sink
2.6 ± 0.5 PgC per year

Fossil fuels and industry
9.0 ± 0.5 PgC per year

Land sink
3.0 ± 0.8 PgC per year

Geological reservoirs

**Figure 11 The carbon budget**

## Do you know?

1 Evaluate the role of volcanoes in the carbon cycle.
2 Outline other forms of natural and human-induced sequestration.
3 Explain how carbon exploitation (such as in power stations) and carbon capture can work together.
4 Outline the link between the global carbon budget and climate change.

# 1.4 Water, carbon, climate and life on Earth

## You need to know

■ the key role that the water and carbon cycles, and their interrelationship, have in supporting life on Earth, with particular reference to climate
■ the role of feedback mechanisms within and between the cycles, and their link to climate change and life on Earth
■ how human interventions in the carbon cycle are designed to influence carbon transfers and mitigate the impacts of climate change
■ how changes in the water and carbon cycles interconnect in the context of a tropical rainforest

# The water and carbon cycles, and climate

Key points:
■ the carbon and water cycles are integral in supporting life on Earth, as they influence climate
■ various feedback mechanisms within and between the cycles are strongly linked to climate change
■ most scientists agree that climate change will significantly impact on life on Earth

The key relationships between the water cycle, carbon cycle and climate change:
■ changes in the carbon cycle are the causes of climate change, through the combination of the greenhouse and enhanced greenhouse effects

## Key terms

**Greenhouse effect** The natural process whereby outgoing thermal radiation is trapped by atmospheric gases such as water vapour and $CO_2$.

**Enhanced greenhouse effect** The increased impact of greater amounts of GHGs ($CO_2$ and methane) caused by human activity.

- climate change is having, and will continue to have, an effect on the water cycle, such as increased evaporation and/or more precipitation in some regions
- climate change is having, and will continue to have, an effect on the carbon cycle, such as the release of more $CO_2$ from permafrost areas as they warm
- humans have to either adapt to the water cycle-related outcomes of climate change (increased rates of ice cap melting, flooding and drought) or mitigate these impacts by managing the carbon cycle, or both

## Feedbacks and climate change

Increased emissions of $CO_2$ are warming the atmosphere through the enhanced greenhouse effect. A number of positive feedback situations have arisen:

- as oceans warm, more water is evaporated, which amplifies natural greenhouse warming
- warm ocean water is less able to absorb $CO_2$, resulting in more $CO_2$ remaining in the atmosphere
- warmer temperatures warm the permafrost, releasing more $CO_2$

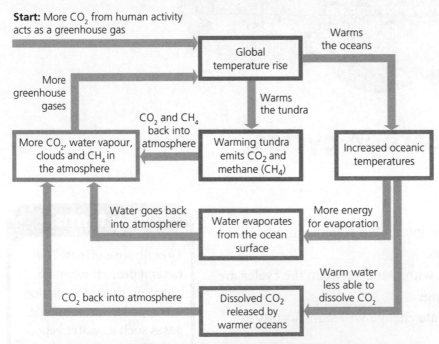

**Figure 12 Positive feedbacks of the links between $CO_2$ warming, evaporation rates and a more moist atmosphere**

# Human interventions in the carbon cycle

Approaches to combating climate change may involve mitigation and adaptation.

**Mitigation:**

- setting targets to reduce GHG emissions
- switching to renewable sources of energy
- 'capturing' carbon emissions and/or storing or burying them (sequestration)

**Adaptation:**

- developing drought-resistant crops
- managing coastline retreat in areas vulnerable to sea-level rise
- investing in fresh water provision to cope with higher levels of drought

## Key terms

**Mitigation** Reduction of the output/amount of greenhouse gases and/or increases in carbon stores (sinks).

**Adaptation** Changing lifestyles to cope with, rather than trying to stop, climate change.

**Anthropogenic** Processes and actions associated with human activity.

## The Kyoto Protocol

The Kyoto Protocol was an intervention in the carbon cycle at a global scale (Table 3). It:

- set legally binding national targets for $CO_2$ emissions compared with 1990
- proposed schemes to enable governments to reach these targets

**Table 3** Evaluation of the Kyoto Protocol

| Successes | Failures |
|---|---|
| Paved the way for new rules and measurements on low carbon legislation, e.g. the UK's Climate Change Act (2008) | Slow ratification — the UK was one of the first to do so; the USA signed the Protocol but did not ratify it; Canada withdrew from it |
| By 2012, carbon emissions in the EU were 22% lower than in 1990 — well ahead of the initial global 5% target | By 2015 there was an increase of 65% in global carbon emissions above 1990 levels, driven by growth in India and China |
| The Clean Development Mechanism supported 75 countries in developing less polluting technology | In order to offset emissions some countries set up complex carbon 'trading' systems, and carbon 'sinks' were allowed |
| Started a global approach to dealing with **anthropogenic** climate change, and more UN conferences on climate change followed | Only industrialised countries were involved, with 'emerging' economies such as India and China left out; the USA's non-ratification has not helped |

## The COP21 Agreement (Paris) 2015

A further conference took place in Paris in late 2015. Three of the main outcomes were:

- temperatures should not increase by more than 1.5°C by 2100
- GHG emissions will be allowed to rise for now, but with more sequestration aimed for later this century to keep within scientifically determined limits

■ emissions targets will be set by countries separately and reviewed every 5 years, with emissions levels decreased meaningfully after each review

**Table 4 An overview of anthropogenic climate change**

| Drivers of anthropogenic climate change | Threats from climate change | Impacts of climate change | Responses to climate change |
|---|---|---|---|
| Fossil fuel burning<br><br>Land-use changes<br><br>Changes to farming practices<br><br>Industrial processes | Changes to the water and carbon cycles<br><br>Shifts in global climates<br><br>Enhanced radiative forcing<br><br>Ocean acidification<br><br>Melting of ice on land and sea<br><br>Ocean thermal expansion<br><br>Rise in sea levels | Alterations to hydrology, such as river flows<br><br>Changes to rainfall and temperature patterns<br><br>Increased river floods, fires and droughts<br><br>Retreat of glaciers<br><br>Increased ocean temperatures — more algal blooms<br><br>Coastal flooding<br><br>Climate change migrants | Mitigation:<br>■ energy efficiency<br>■ energy conservation<br>■ alternative energy<br>■ carbon-trading systems<br>■ emission taxes<br><br>Adaptation:<br>■ resource management<br>■ migration corridors<br>■ farming of regionally appropriate crops/products<br>■ shoreline protection |

# The Amazon rainforest
## The water cycle and climate change

Key points:
■ an increase in atmospheric temperature will increase rates of evapotranspiration. Sea temperatures are expected to warm, particularly in the Pacific Ocean. The ENSO may occur more frequently
■ inland, a decrease in precipitation during the dry season. Reduced rainfall and prolonged drought are features of an ENSO year. More intense rainfall during the wet season
■ sea levels are currently rising by 5 mm per annum within the Amazon delta. Increased erosion and flooding is likely to have a substantial impact on low-lying areas and will destroy the coastal mangrove forests
■ glaciers in the Andes provide the source for 50% of the discharge of the upper Amazon. Over the last 30 years, Peruvian glaciers have shrunk by 20% and it is predicted that Peru will lose all glaciers below a height of 6000 m by 2050

> ### Key term
>
> **ENSO** The El Niño Southern Oscillation — a warm ocean current that replaces the usual cold current off the Pacific coast of South America. It brings heavier rain than usual on the coast, and drought inland.

# The carbon cycle and climate change

Key points:

■ 40% of plant species may become unviable in the Amazon by 2050. Large areas of the tropical rainforest may be succeeded by mixed forest and savanna grassland, creating a reduction in the net carbon store of the rainforest

■ as the dry season lengthens, trees will have more time to dry out so there will be an increased incidence of wildfires. $CO_2$ emissions will increase

■ forest die-back and wildfires are predicted to result in the Amazon region becoming a net source of $CO_2$, rather than a carbon sink

■ Amazonian soils under rainforest contain up to $9\,kg\,m^{-2}$ of carbon in the upper 50 cm of the soil profile. Clearance of the rainforest will result in less carbon stored in the soil

## Do you know?

1 Explain how the enhanced greenhouse effect operates.

2 Summarise the key features of the Kyoto Protocol.

3 Describe and explain how one carbon trading system operates.

4 Research two other outcomes of the COP21 (Paris) Agreement.

5 Outline the impacts of the ENSO on South America.

## End of section 1 questions

1 What is meant by (a) the 'water balance' and (b) the carbon budget?

2 Explain how land-use change can affect the water cycle.

3 With reference to a river catchment that you have studied, examine the impact of precipitation on drainage basin stores and transfers.

4 With reference to a river catchment that you have studied, assess the potential impact of human activity on the drainage basin.

5 Outline the process of photosynthesis in the carbon cycle.

6 How far do you agree that changes to the carbon cycle will lead to increasingly severe storm events?

7 Assess the relative importance of natural factors in changing the size of major stores of carbon.

8 In the context of climate change, distinguish between mitigation and adaptation.

9 'Human activity has caused irreversible damage to the fragile interrelationship between the water cycle and the carbon cycle.' To what extent do you agree with this view?

10 To what extent do you agree that human activity is responsible for permanent changes to the carbon cycle in tropical rainforests?

# 2 Coastal systems and landscapes

## 2.1 Coasts as natural systems

### You need to know

- how systems apply to coastal landscapes
- how feedback mechanisms operate in coastal areas
- the distinction between landform and landscape

## Coastal landscapes as systems

Coastal environments act as natural open systems (Table 5).

Table 5 Features of a coastal system

| Inputs | Weathering/ erosional processes | Erosional components | Transport processes (flows) | Depositional components (stores) | Outputs |
|---|---|---|---|---|---|
| Marine:<br>■ energy from waves<br>■ tides and sea currents<br>■ salt spray<br><br>Geological:<br>■ rock type<br>■ rock structure<br>■ products of weathering<br><br>Atmospheric:<br>■ wind energy<br>■ precipitation<br>■ temperature<br><br>Sea-level change<br><br>Human activity:<br>■ land use<br>■ coastal protection | Weathering:<br>■ physical<br>■ chemical<br>■ biological<br><br>Erosion:<br>■ hydraulic action<br>■ wave quarrying<br>■ abrasion<br>■ attrition<br><br>Mass movement:<br>■ landslides<br>■ rockfalls<br>■ mudflows<br>■ rotational slips<br>■ soil creep | Erosional landforms and landscapes:<br>■ cliffs<br>■ headlands and bays<br>■ wave-cut platforms<br>■ geos<br>■ caves, arches and stacks | Water transport:<br>■ longshore drift<br>■ onshore and offshore movement<br>■ traction<br>■ saltation<br>■ suspension<br><br>Wind transport:<br>■ surface creep<br>■ saltation | Depositional landforms and landscapes:<br>■ beaches<br>■ spits<br>■ tombolos<br>■ bars and barrier beaches<br>■ sand dunes<br>■ salt marshes | Energy<br><br>Onshore sediment<br><br>Marine sediment |

# Feedback

Feedback mechanisms operate in coastal areas.

**Positive feedback:**

- sea walls may prevent flooding, but they also limit cliff erosion
- this restricts the release of sediment into the system
- this sediment might otherwise have been redeposited, so helping to protect the coastline

**Negative feedback:**

- sediment eroded from a beach during a storm is then deposited offshore to form a bar
- waves break before reaching the beach, dissipating their energy and therefore reducing erosion of the beach
- normal wave conditions rework offshore deposits back to the beach

# Landscape vs landform

Key features:

- as a result of feedback mechanisms and constantly changing weather conditions, dynamic equilibrium is rare in a coastal landscape
- coastal landscapes are made up of a combination of wide-ranging erosional and depositional landforms, which are constantly undergoing change. They may also be affected by human activities, including management

## Key terms

**Positive feedback** Where a change causes a further (or snowball) effect, continuing or even accelerating the original change.

**Negative feedback** Lessens the effect of the original change and ultimately reverses it.

**Landscape** An expanse of land/scenery that can be seen in a single view. It covers all aspects of the view, both natural landforms and human-created features.

**Landform** A single natural feature (such as a cliff or beach) found in a landscape.

## Do you know?

1 With reference to one coastal area you have studied, identify the main stores in that area.

2 Identify one feedback mechanism arising from human activity in a coastal area.

# 2.2 Systems and processes

**You need to know**
- the main sources of energy in a coastal environment
- the concepts of sediment cells and sediment budgets
- the processes that operate in a coastal environment, both geomorphological and marine

## Sources of energy

### Wind and waves

Wave energy is controlled by the:
- force of the wind and its direction
- duration of the wind
- **fetch** — the longer the fetch, the more energy that waves possess

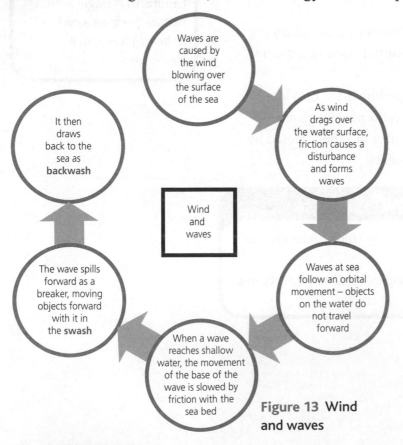

**Figure 13** Wind and waves

### Key terms

**Fetch** The sea distance over which the wind has blown to produce waves.

**Swash** The landward flow of water up a beach.

**Backwash** The seaward flow of water down a beach.

Constructive waves:
- build beaches
- have longer wavelengths, lower height and are less frequent (at 6–8 per minute)
- swash is greater than backwash so it adds to beach materials, giving rise to a gently sloping beach

- the upper part of such a beach is marked by a series of small ridges called berms, each representing the highest point the waves have reached at a previous high tide

Destructive waves:

- have shorter wavelengths, greater height and are more frequent (at 10–14 per minute)
- backwash is greater than the swash so that sediment is dragged offshore
- create a steeper beach profile initially, though over time the beach will flatten as material is drawn backwards
- form shingle ridges at the back of a beach (storm beaches), created by storms

## Currents

Strong underwater currents can be present offshore:

- they create alternating shallow and deep sections (ridges and runnels)
- the channels created can run either parallel to the beach or at right angles to it
- strong rip currents can be present in these channels and can be dangerous

## High-energy and low-energy coasts

High energy:

- occur where wave power is strong for a greater part of the year
- in the UK, they occur mostly on west-facing coasts — the direction of the longest fetch
- waves up to 30 m have been recorded on the west coast of Ireland

Low energy:

- represented by estuaries, inlets and sheltered bays — areas of deposition
- wave heights are considerably lower than on high-energy coasts
- enclosed seas, such as the Baltic Sea, also contain low-energy environments

## Sediment cells and budgets

Sediment sources:

- rivers
- the seabed
- erosion of the coastline
- shell material

Sediment cells:

- DEFRA has identified eleven major sediment cells for England and Wales as basic units for coastal management

- each cell is separated by headlands or stretches of open water
- most cells are divided into sub-cells
- sediment cells are a key component of shoreline management plans, which decide on future strategies for coastal management (see page 40)

Sediment budgets:

- see Figure 14

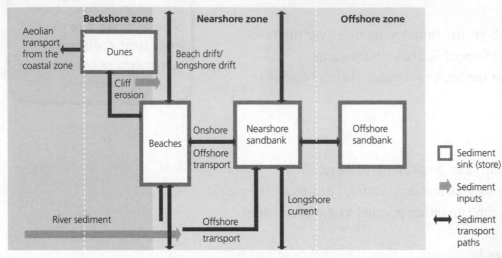

**Figure 14** Stores and flows in coastal sediment budgets

# Geomorphological processes

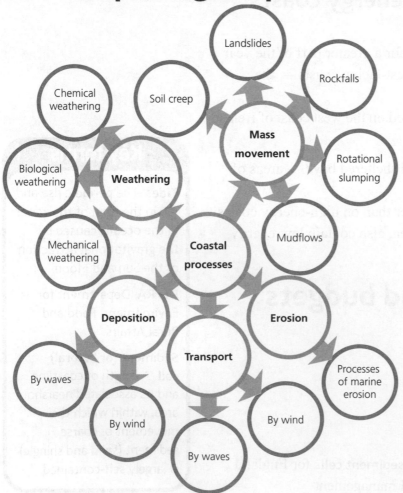

**Figure 15** Coastal geomorphological processes

**Key term**

Sediment budget The relationship between accretion and erosion, which can be used to predict the changing shape of a coastline over time.

# Weathering

Physical weathering:

■ involves the breakdown of rocks into smaller fragments through mechanical processes such as expansion and contraction, mainly due to temperature change

■ frost shattering and salt weathering are common on coasts — they are examples of **sub-aerial weathering**

Frost shattering:

■ see Figure 16

**Key terms**

**Sub-aerial weathering** The collective name for weathering processes on the Earth's surface (literally at the base of the atmosphere).

**Scree** Collections of loose rock at the base of a slope.

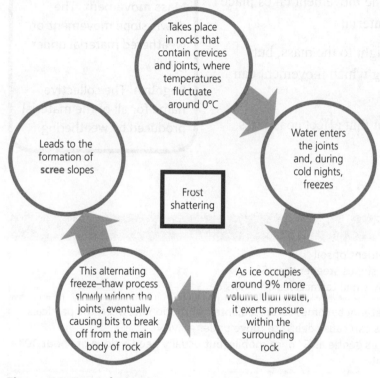

**Figure 16 Frost shattering**

Salt weathering:

■ takes place when a rock becomes saturated with water that contains salt

■ some of the salt crystallises and begins to exert pressure on the rock as the salt crystals are larger than the spaces in which they being formed

■ the process repeats over time, and causes the disintegration of rock

Chemical weathering:

■ involves the decay or decomposition of rock in situ

■ usually takes place in the presence of water, which acts as a dilute acid

■ the end-products are either soluble and removed in solution, or have a larger volume than the mineral they replace

■ tends to increase with rising temperature and humidity levels

**Exam tip**

Ensure you know the difference between weathering and erosional processes.

- for carbonic acid (carbonation) on limestones, however, lower temperatures produce greater rates of weathering
- can also occur from the action of dilute acids resulting from both atmospheric pollution (sulfuric acid), and the decay of plants and animals (organic acids)

# Mass movement

The rate of **mass movement** depends on:
- the degree of cohesion of the **regolith**
- the steepness of the slope down which the movement takes place
- the amount of water contained in the material

Note that a large amount of water adds weight to the mass, but more importantly lubricates the plane along which movement can take place.

There are various forms of mass movement that affect coastlines (Table 6).

## Key terms

**Mass movement** The down-slope movement of weathered material under gravity.

**Regolith** The collective name for all of the material produced by weathering.

**Table 6 Types of mass movement on coasts**

| Mass movement | Commentary |
|---|---|
| Creep | ■ slow downhill movement of soil<br>■ tends to operate on slopes steeper than 6°<br>■ evidence is shown by small terracettes on a hillside |
| Earthflows | ■ when weathered material becomes saturated, internal friction between the particles is reduced and stress can cause debris to move under gravity<br>■ can occur on slopes as gentle as 5° once mobile, but usually need a slope of about 10° to initiate movement |
| Mudflows | ■ more rapid than earthflows<br>■ occur in areas that experience torrential rain falling on ground that has limited protection from vegetation cover<br>■ regolith becomes saturated, increases the water pressure in the debris and reduces the frictional resistance between particles |
| Rock falls | ■ where erosion is concentrated at the base of a cliff, it becomes unstable and collapses into the sea |
| Landslides | ■ occur when rocks and/or regolith have bedding planes or when material in one layer becomes very wet and over-lubricated<br>■ the added weight from water causes the plane/layer to slip under gravity downwards over the underlying layers |
| Slumping | ■ saturated material moves suddenly, resulting in whole sections of cliff moving downwards<br>■ often happens where softer material overlies strata that are far more resistant<br>■ the slip plane is often concave, producing a rotational movement |

# Marine-based processes
## Erosion

Coastal erosion operates by a variety of processes (Table 7).

**Table 7 Coastal erosion processes**

| Erosion process | Commentary |
| --- | --- |
| Hydraulic action | ■ where a wave breaking against rocks traps air into cracks under pressure, which is then released suddenly as the wave retreats<br>■ causes stress that creates more cracks, allowing the rock to break up more easily<br>■ includes pounding — the sheer weight and force of water pushing against a cliff face causing it to weaken |
| Abrasion (corrasion) | ■ where material carried by waves (the load) is used as ammunition to wear away rocks on a cliff or a wave-cut platform as the material is thrown or rubbed against them repeatedly by each wave<br>■ abrasion targeted at specific areas such as notches or caves is known as quarrying |
| Attrition | ■ loose rocks are broken down into smaller and more rounded pebbles, which are then used in abrasion |
| Cavitation | ■ when air bubbles trapped in fast-moving water collapse, causing shock waves to break against the rocks under the water<br>■ repeated shocks weaken the rock |

The rate of coastal erosion is governed by several factors:

■ geology — harder rock (e.g. granite) is more difficult to erode than softer material (e.g. boulder clay). Some more resistant rock can be eroded along joints or cracks. Limestones are prone to weathering

■ structure and dip of rocks — if rocks dip inland or are horizontally bedded, steep cliffs form; rocks that dip seaward produce gentler slopes

■ coastal shape — softer rock is eroded to form bays, with harder rocks forming headlands

■ wave refraction — erosion is concentrated on headlands, whereas in bays waves spread over a wider area and their energy is dissipated, causing more deposition

■ wave steepness — steeper high-energy waves have more power to erode

■ wave breaking point — waves breaking at the foot of a cliff have more energy to erode

■ fetch — waves that have travelled a long distance have more energy

■ width of beaches — beaches absorb some of the waves' energy and protect the coastline

■ nature of beaches — a pebble beach dissipates energy from waves through friction and percolation

■ human developments:
  □ sand and pebble extraction from beaches for use as building materials weakens the coastline and can lead to erosion

> ### Exam tip
>
> Despite what it may say in the specification, solution is *not* a major form of erosion by the sea — the water is too alkaline.

☐ sea walls, groynes and other coastal protection schemes help protect an area from erosion, but can also increase erosion further along the coast

☐ development on cliff tops can increase runoff and cause instability and cliff failure

## Transportation

Material is transported along a coastline by:

- swash and backwash (see page 26)
- longshore (littoral) drift — material is moved along a shoreline by waves that approach at an angle. Swash moves sand and shingle up the beach at an angle but the backwash is at right angles to the beach. This results in material zigzagging its way along the beach according to the prevailing wave direction. Both material in **suspension** and larger fragments are moved this way
- traction — when large stones and pebbles are rolled along the seabed and across a beach

Material is also transported in coastal areas by runoff — the flow of water overland, either in small channels (rills) or as streams and rivers.

## Deposition

Deposition occurs in low-energy environments, such as bays and estuaries:

- when sand is deposited on a beach and dries out, it can be blown by the wind further inland to form sand dunes at the back of the beach
- in a river estuary, mud and silt can build up in sheltered water to create a salt marsh. Here the fresh water of the river meets the salt water of the sea, causing **flocculation** of suspended material to occur, creating extensive areas of mudflats

### Key terms

**Suspension** The process by which very small particles are held in water.

**Flocculation** The process by which a river's load of clays and silts carried in suspension is deposited more easily on its meeting with sodium chloride in sea water.

### Do you know?

1 What causes a wave to break?
2 Describe the nature of rip currents.
3 Distinguish between mechanical weathering and chemical weathering.
4 What are the key differences between a flow, a slide and a slump?
5 Explain the process of wave refraction.

# 2.3 Coastal landscape development

## You need to know

- the characteristics and development of landforms/ landscapes associated with coastal erosion, coastal deposition, estuarine conditions and sea-level change
- the impact of climate change on coasts

## Erosional landscapes

A coastal erosional landscape will include all or some of the following.

Cliffs, headlands and bays:

- form when rocks of differing hardnesses are exposed together at a coastline
- tougher, more resistant rocks (such as granite and limestones) tend to form headlands with cliffs
- weaker rocks (such as clays and shales) are eroded to form sandy bays

Wave-cut platforms:

- see Figure 17

**Exam tip**

Landscapes are the outcome of processes, so you will also need to refer to topic 2.2 Systems and processes here.

**Exam tip**

Make sure that you refer to examples from both the UK and other countries.

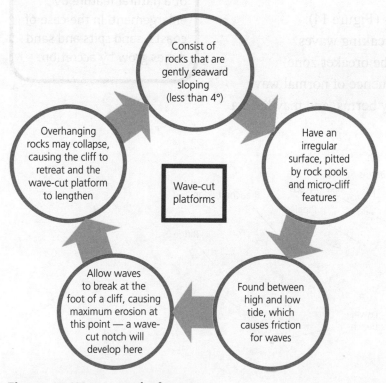

**Figure 17 Wave-cut platforms**

(Diagram contents:)

- Consist of rocks that are gently seaward sloping (less than 4°)
- Have an irregular surface, pitted by rock pools and micro-cliff features
- **Wave-cut platforms**
- Found between high and low tide, which causes friction for waves
- Allow waves to break at the foot of a cliff, causing maximum erosion at this point — a wave-cut notch will develop here
- Overhanging rocks may collapse, causing the cliff to retreat and the wave-cut platform to lengthen

Caves, arches and stacks:

- erosion takes place on a cliff face where there is a weakness, such as joints or bedding planes
- where waves open up a prolonged joint, they form a deep and steep-sided inlet (geo)
- smaller hollows can be excavated to create caves
- where caves are created on either side of a headland and are eroded back, they can 'meet' each other (the back wall collapses) and form an arch
- the sea is now able to splash under the arch, further weakening it until eventually the roof collapses, leaving the seaward side as a separate island (a stack)
- over time the stack erodes to form a stump

# Depositional landscapes

A coastal depositional landscape will include all or some of the following.

Beaches:

- built up by **accretion** in and across bays and made of either sand or shingle, or a mixture of both
- are either
  - □ swash-aligned, where sediment is taken up and down the beach with little sideways transfer
  - □ drift-aligned, where sediment is transferred along a beach by longshore drift
- can be sub-divided into different zones (Figure 14):
  - □ offshore, beyond the influence of breaking waves
  - □ nearshore — intertidal and within the breaker zone
  - □ backshore — usually above the influence of normal wave patterns, marked at the lower end by berms, and may have a storm beach further up (Figure 18)

## Exam tip

A good way to demonstrate that you know what each of these erosional and depositional landforms looks like is to draw a sketch. Give it a go!

## Key term

**Accretion** The growth of a natural feature by enlargement. In the case of coasts, sand spits and sand dunes grow by accretion.

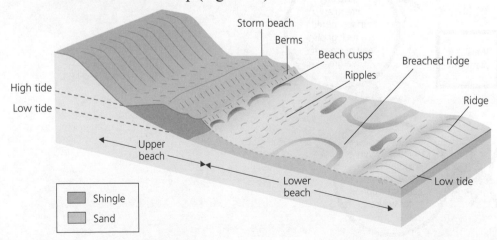

**Figure 18 Beach profile features**

Small-scale beach landforms:

■ ridges and runnels — alternate raised and dip sections that run parallel to the shore line, exposed at low tide but hidden at high tide. They are caused by strong backwash, and strong tides

■ beach cusps — small semi-circular depressions in sand and shingle beaches. Once created they self-perpetuate, especially on swash-aligned beaches

■ ripples — micro beach ridges parallel to the shoreline, created by wave action on low-gradient beaches

Spits, tombolos:

■ long, narrow stretches of sand or shingle that protrude into the sea or across an estuary

■ result from materials being moved along the coast by longshore drift

■ this movement continues in the same direction when the coastline curves; where there is an estuary with a strong current that interrupts the movement of material, they project out into it

■ the end of the spit is often curved (creating a series of laterals) where waves are refracted around the end of the spit into more sheltered water behind

■ a tombolo is formed where a spit joins the mainland at one end to an island at the other

Bars, barrier beaches:

■ created where a spit develops across a bay because there are no strong currents to disturb the process — the water behind it is dammed, forming a lagoon

■ bars also develop as a result of storms raking up pebbles. This shingle left in offshore ridges creates a barrier beach

Offshore bars:

■ deposits of sand and shingle situated some distance from a coastline — these usually lie below sea level, becoming visible only at low tide

■ there are two explanations as to where and how they form:
  □ in shallow seas where the waves break some distance from the shore
  □ where steep waves break on a beach, creating a strong backwash that carries material back down the beach to form a ridge

■ when a bar appears above the level of the sea for most of the time it becomes a barrier beach, with a lagoon on the landward side and ocean on the other

### Exam tip

You may be asked to describe any of these landforms. Refer to size, shape, nature of sediments and field relationship (where the landform lies in relation to the landscape).

Sand dunes:

■ sand is often deposited by the sea under low-energy conditions
■ wind may then move the sand to build dunes further up the beach
■ these in turn become colonised by stabilising plants (a **psammosere**)

# Estuarine environments

Key features:

■ sheltered river estuaries or zones in the lee of spits are areas where there are extensive accumulations of silt and mud (mudflats), aided by flocculation and gentle tides
■ these inter-tidal areas are colonised by vegetation, and a succession of plant types may develop over time (a **halosere**), creating a salt marsh (Figure 19)
■ the initial plants of a halosere are tolerant of both salt and regular inundation at high tide
■ they also have a long root system and a mat of surface roots to hold the mud in place
■ the plants trap more mud and build up a soil for the next stage of the succession
■ as the mat of vegetation becomes more dense, the impact of the tidal currents reduces and humus levels increase, allowing reeds and rushes to grow and, later, alder, willow and oak
■ salt marshes often have complex systems of waterways — creeks
■ in some extensive salt marsh areas, hollows of trapped sea water form, which then evaporate and create salt pans

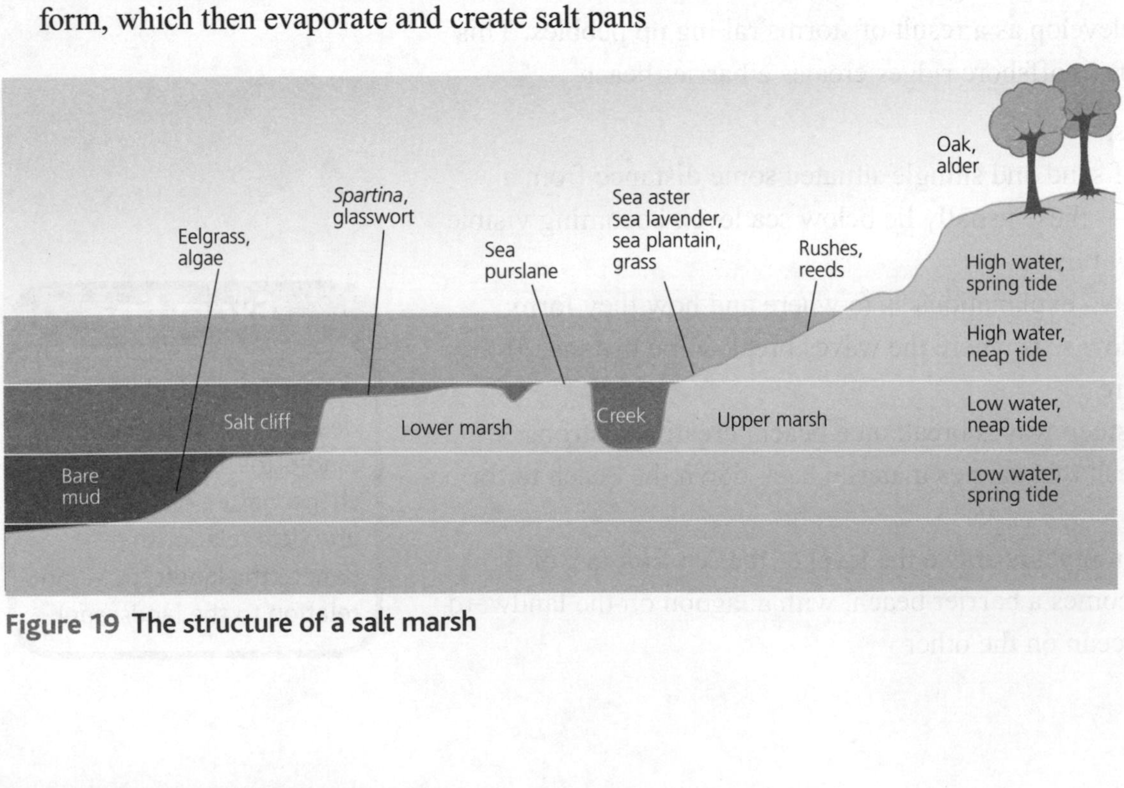

> ## Key terms
>
> **Psammosere** The succession of plants that develops on a sand dune complex. Plants include sea rocket and lyme grass nearer the sea, with marram grass, fescue and gorse inland.
>
> **Halosere** The succession of plants that develops in a salt marsh. Plants include eelgrass, *Spartina*, cord grass and sea lavender.

**Figure 19 The structure of a salt marsh**

# Landscapes of sea-level change

Changes in sea level take place over time due to:

■ sea temperatures being colder or warmer than the present
■ relative changes in land levels

## Eustatic change

Key features:

■ results from a fall in sea level due to a new glacial period, when water is held as ice
■ results from a rise in sea level when, at the end of a glacial period, the ice on land melts

## Isostatic change

Key points:

■ arises from changes in the local relationship of land to sea
■ as ice collected on the land during a glacial period, the extra weight pressed down on the land, causing it to sink and sea level to rise
■ as the land ice melts, the land begins to move back up to its original position (readjustment) and sea level falls
■ depends on the thickness of the original ice and the speed of its melting

> ### Key terms
>
> **Eustatic change** A sea-level change that affects the globe.
>
> **Isostatic change** A sea-level change that affects a localised area.

## Tectonic change

Other changes of sea level have been caused by tectonic processes associated with plate movement — they too tend to be localised.

## Major changes in sea level

In the last 10,000 years (the Holocene):

■ global sea level rose very quickly up to 6000 years ago (the Flandrian transgression)
■ it flooded the North Sea, English Channel and Irish Sea
■ it flooded many former river valleys to give the distinctive indented coastline of southwest England and Ireland (rias)
■ since then sea levels have remained largely consistent, with a slight rise recently due to climate change

# Coastlines of emergence and submergence

## Coastlines of submergence

Rias:

■ drowned winding river valleys with long fingers of water stretching a long way inland, including their tributary valleys

■ are widest and deepest nearest to the sea and get progressively narrower and shallower inland

■ tidal changes will often reveal extensive areas of mudflats

Fjords:

■ straight, glaciated valleys that have been drowned by rising sea levels at the end of the ice ages

■ have a shallower area at the mouth (a rock threshold), where the ice thinned as it reached the sea and hence lost its erosional power

■ have the typical steep-sided and deep cross profile associated with glacial troughs, and can stretch many kilometres inland

Dalmation coasts:

■ a drowned coastline where the main relief trends run parallel with the line of the coast

■ ridges of upland produce elongated islands separated from the mainland by flooded valley areas

■ their name originates from the Adriatic coast of Dalmatia (Croatia)

## Coastlines of emergence

The effect of falling sea levels is to expose land normally covered by the sea:

■ cliffs that are no longer being eroded become isolated from the sea, leaving relic cliffs

■ 'fossil' features, such as former caves and stacks, are left higher up from the coast on raised marine platforms

■ raised beaches — common on the coast of western Scotland, where a series of raised sandy and pebble-ridden terraces can be found above the current sea levels

# Climate change

Sea-level rise associated with climate change is important as increases of several centimetres are predicted in the coming decades, due to:

■ the thermal expansion of water as it becomes warmer

### Exam tip

Be aware of the links between process and landform and between landforms and landscapes, and of how changes in time may influence both landforms and landscapes.

■ more water being added to the oceans following the melting of freshwater glaciers and ice sheets, such as those in Greenland

## Do you know?

1 Explain why wave-cut platforms tend to have a maximum width of about 0.5 km.

2 Why can waves erode both sides of a headland?

3 Give two reasons why temperature change causes sea-level change.

4 Describe the impact of isostatic sea-level change on the British Isles.

5 Make three lists: coastal landforms created by erosion; coastal landforms created by deposition; and coastal landforms resulting from sea-level change.

# 2.4 Coastal management

## You need to know

■ the traditional approaches to coastal management — 'hard' and 'soft' engineering

■ more sustainable approaches to managing coastal flood risk and erosion: the principles of shoreline management plans (SMPs)

## Key terms

**Hard engineering** A form of coastal management that involves the construction of man-made features.

**Soft engineering** A form of coastal management that involves working with nature and natural features.

**Integrated approach** A combination of hard and soft engineering.

## Hard and soft engineering

Traditionally, coastal defence strategies against the risks of coastal flood and erosion are classified into 'hard engineering' methods (Table 8) and 'soft engineering' methods (Table 9). Many modern coastal management schemes have an integrated approach.

Table 8 Hard engineering strategies

| Strategy | Description | Commentary |
|---|---|---|
| Sea walls | Concrete or rock walls at the foot of a cliff or at the top of a beach; usually have a curved face to reflect waves back out to sea | Although often effective at the location where they are built, they deflect erosion further along the coast; they are expensive and have high maintenance costs |
| Groynes | Timber or rock structures, built at right angles to the coastline; trap sediment being moved along the coast by longshore drift | The beach created increases tourist potential, and gives protection to the land behind; the process starves beaches further down the coast of sand, however, increasing erosion there |
| Rip-rap (rock armour) | Large, hard rocks dumped at the base of a cliff or at the top of a beach; forms a permeable barrier to breaking sea waves | Relatively cheap, and easy to construct and maintain; the rocks used are often brought in from other areas and hence may not blend in |
| Revetments | Wooden barriers, in a slat-like form, placed at the base of a cliff or top of a beach | Intrusive and very unnatural |
| Gabions | Wire cages filled with small rocks that are built up to make walls; often used to support weak cliffs | Relatively inexpensive; look unsightly to begin with but as vegetation grows they blend in; the metal cages rust and break easily |

**Table 9 Soft engineering strategies**

| Strategy | Description | Commentary |
|---|---|---|
| Beach nourishment | Addition of sand or pebbles to an existing beach to make it higher or wider; the materials are usually dredged from the nearby seabed and spread or 'sprayed' on to the beach | A relatively cheap and easy process; the materials used blend into the natural beach; it is a constant requirement, however, as natural processes may continue to move materials away |
| Dune regeneration | Planting of marram grass and other plants that bind sand together; areas are often fenced off to keep people off newly planted dunes | Maintains the look of a natural coastline and provides important habitats; process requires a lot of time to be effective |
| Marsh creation | Low-lying coastal lands are allowed to be flooded by the sea; the area becomes a salt marsh | Provides an effective buffer to the power of waves, creating a natural defence; creates an opportunity for wildlife habitats; agricultural land is lost, however, and landowners require compensation |

# Shoreline management plans

Key features:

- SMPs were introduced in 1995, with 22 in England and Wales
- do not exist in Scotland and Northern Ireland, where the devolved governments and local authorities are jointly responsible for coastal protection
- involve all stakeholders in making decisions about how coastal erosion and coastal flood risk should be managed
- aim to balance economic, social and environmental needs and pressures at the coast
- reduce risks to people and to the developed, historic and natural environment in a sustainable way
- predict, so far as it is possible, the way in which a coastline will be shaped in the future (defined as 100 years)

Within the work of SMPs, four policies are often considered (Figure 20).

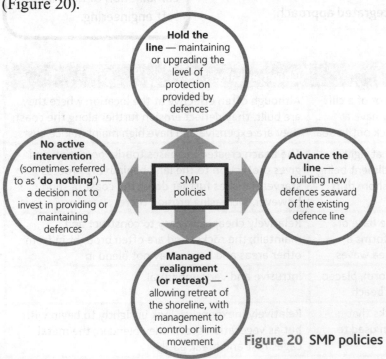

**Figure 20 SMP policies**

## Do you know?

1 Which type of hard engineering interferes most with longshore drift?

2 Compare hard and soft engineering in general terms.

3 Who manages shoreline management plans (SMPs), and why is this important?

4 Which coastal zone management policy option involves 'letting nature take its course'?

## End of section 2 questions

1 Outline the role of wind in affecting coastal energy.

2 Outline how the coast is described as a natural system.

3 Explain how tides are created.

4 Distinguish between eustatic and isostatic sea-level change.

5 Assess the importance of different sources of energy in the creation of coastal landscapes.

6 Evaluate the role of sea-level change over the last 10,000 years in the development of coastal landscapes.

7 Explain how soft engineering could protect a coastline.

8 Hard engineering has been used to protect some coasts. With reference to a case study, explain how hard engineering can protect the coast and comment on its effectiveness.

9 'Coastal flooding and erosion will become a more common occurrence over the coming decades.' To what extent do you agree with this view?

10 Assess the extent to which predicted climate change will present challenges for the sustainable management of a local-scale coastal environment that you have studied.

# 3 Hazards

## 3.1 The concept of hazard in a geographical context

### You need to know

- the nature, forms and potential impacts of natural hazards
- the variety of perceptions of, and responses to, hazards
- the relationships between responses and hazard incidence, intensity, magnitude, distribution and level of development
- theories of hazard response (Park model, hazard management cycle)

## Types of hazard and their impact

There are various types of **hazard**:

- geophysical — volcanoes, earthquakes and tsunamis
- hydrological — droughts and floods
- atmospheric — tropical storms, tornadoes and extra-tropical storms (deep depressions)
- biohazards — wildfires and locust plagues

Key points:

- few hazards are entirely natural
- their impact and relationship with **disaster** are the result of human **vulnerability**
- human actions can intensify the impact of natural hazards — e.g. exacerbating earthquake **risk** by building inappropriate buildings

The hazard risk equation helps to make clear the relationship between hazards and disasters:

$$\text{risk} = \frac{\text{hazard} \times \text{vulnerability}}{\text{capacity to cope}}$$

Some communities have a high **resilience** to hazards. They can reduce the chances of disasters occurring because:

- they have emergency evacuation, rescue and relief systems in place
- they react by helping each other, reducing numbers affected
- the use of hazard-resistant design or land-use planning has reduced the numbers at risk

### Key terms

**Hazard** A natural/geophysical event that has the potential to threaten both life and property.

**Disaster** When a hazard has a significant impact on people — a realisation of a hazard.

**Vulnerability** The risk of exposure to hazards combined with an inability to cope with them.

**Risk** The probability of a hazard occurring and leading to a loss of lives and/or livelihoods.

**Resilience** The degree to which a population or environment can absorb a hazardous event and yet remain within the same state of organisation — its ability to cope with stress and recover.

# Perception and responses

## Perception

People perceive the threat of hazards in different ways. Perception is influenced by:

- socio-economic status — including employment status
- level of education and income
- religion
- ethnicity
- family situation
- past experience
- values, attitudes and expectations

In general:

- wealthier, more educated people may perceive that they are better prepared and may have more government support
- a sense of helplessness, or **fatalism**, tends to increase as poverty levels increase (note: this feeling can also apply to more disadvantaged people within a wealthy country)

## Responses

Perception of a hazard determines the course of action taken by individuals in order to modify the event and the responses they expect from governments and other organisations.

Responses to hazards include:

- fatalism — some communities believe that hazards are 'God's will'
- **adjustment/adaptation** — people see that they can prepare for, and therefore survive, the event(s) by **prediction**, prevention and/or protection, depending on the economic and technological circumstances of the area affected
- **mitigation** — ranging from monitoring, education and community awareness of what to do (**risk sharing**), to various technological strategies for shock-proof building design (e.g. in Tokyo and San Francisco) or protection (e.g. Japanese tsunami walls)
- efficient management — involves information gathering, analysis and planning; this is linked to levels of governance

### Key terms

**Fatalism** An acceptance that hazards are natural events and are part of living in an area.

**Adjustment/adaptation** The changing of lifestyles or behaviours to cope with the threats and impacts before and after a hazardous event.

**Prediction** The ability to give warnings so that action can be taken to reduce the impact of hazard events. Improved monitoring and use of ICT have meant that predicting hazards and issuing warnings has become more effective.

**Mitigation** The reduction of the amount and scale of threat and damage caused by a hazardous event.

**Risk sharing** The ways in which communities come together to reduce loss of life and damage.

## Factors influencing responses

Incidence:

■ how often an event occurs, sometimes called the recurrence interval, such as 'a 1-in-100-year event'

■ for most hazards there is usually an inverse relationship between incidence and magnitude

Intensity:

■ refers to the magnitude of a tropical storm — related to the level of atmospheric pressure within the storm

■ measured by the Saffir-Simpson scale

Magnitude:

■ size of an event (earthquake or volcano)

■ correlation between magnitude and the level of disaster is far from direct

■ earthquake magnitude is measured by the logarithmic moment magnitude scale (MMS), and the damaging effects are measured by the Mercalli scale

■ volcano magnitude is measured by the volcanic explosivity index (VEI), based on the volume and column height of ejections

Distribution:

■ areal extent of a hazard

■ can have a large impact, e.g. the Icelandic ash clouds after the Eyjafjallajökull eruption (2012)

Level of economic development:

■ vulnerability is closely associated with levels of absolute poverty and inequality

■ the poorest countries lack money to invest in education, social services, basic infrastructure and technology, all of which help communities overcome disasters

■ economic growth increases economic assets and therefore raises potential risk levels unless managed effectively

### Exam tip

Note how this section and the following section interconnect with other parts of the specification — globalisation, governance, urbanisation, population and the character of places. Some questions will require you to make these connections and links.

# Hazard response models
## The Park model

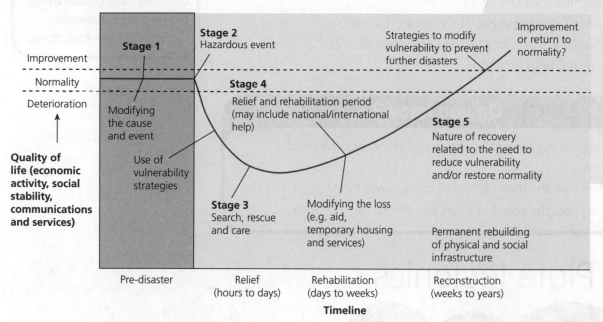

**Figure 21** **The Park disaster response model**

This model illustrates:

- how quality of life is impacted by a hazardous event
- how a range of management strategies can be used over time — from before the event to after the event
- the roles of emergency relief agencies and rehabilitation
- that different areas affected may have different response curves, depending on level of preparedness and economic development

**Exam tip**

Be prepared to apply Park's model to each of the case studies you examine for volcanoes, earthquakes, tropical storms and wildfires. Try to look for both similarities and differences in all of these events.

## The hazard management cycle

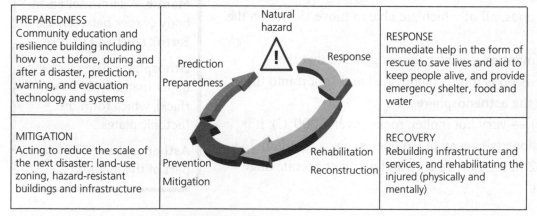

**Figure 22** **The hazard management cycle**

This illustrates four stages of the management of hazards in seeking to reduce the scale of a disaster. Note the recovery ('returning to normal') stage depends on:

- the magnitude of the event
- the level of development of the area affected
- the level of governance of the area affected
- the external help available

**Exam tip**

Be prepared to analyse and compare case studies using the hazard management cycle and/or the Park model.

## Do you know?

1 Distinguish between the primary and secondary effects of a hazard.
2 Explain how the level of risk can change over time.
3 Explain how the speed of onset of a hazard can be critical.

# 3.2 Plate tectonics

## You need to know

- the structure of the Earth and plate tectonics theory
- processes and landforms at different plate margins
- magma plumes and their relationship to plate movement

## Earth structure

The Earth's structure (Figure 23) consists of:

- an outer layer (crust) split up into seven large rigid plates and several smaller ones, all of which are able to move slowly on the surface
- a middle layer (**mantle**) — some have suggested that the crust/mantle division is more complicated and split it into the **lithosphere** and the **asthenosphere**
- the centre (core) — very hot molten rocks (over 5,000°C): it is thought that radioactive decay of isotopes such as uranium-238 and thorium-232 in the Earth's core and mantle generate huge amounts of heat

**Key terms**

**Mantle** A semi-molten body of rock between the Earth's crust and its core.

**Lithosphere** The crust and upper mantle (80–90 km thick), which form the tectonic plates.

**Asthenosphere** The lower part of the mantle.

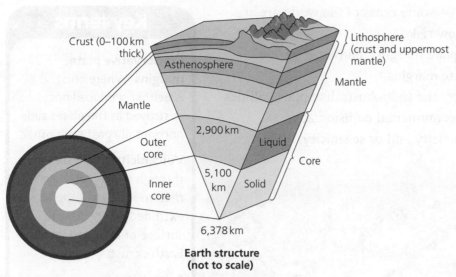

**Figure 23** The structure of the Earth

# Plate tectonics theory

Various theories/discoveries have been developed/achieved to support the concept of plate tectonics:

- Wegener's continental drift hypothesis (1912) postulated that now-separate continents had once been joined together
- in the 1930s it was suggested that Earth's internal radioactive heat was the driving force of convection currents in the mantle that could move tectonic plates on the surface
- the discovery (1960s) of magnetic stripes in the oceanic crust of the seabed — these are palaeomagnetic (ancient magnetism) signals from past reversals of the Earth's magnetic field and prove that new crust is created by the process of sea-floor spreading at mid-ocean ridges
- at **constructive (divergent) plate margins**, elevated altitudes of oceanic crust at ridges create a 'slope', down which oceanic plates slide (gravitational sliding or ridge push)
- at **destructive (convergent) plate margins**, high-density ocean floor is being dragged down by a downward gravitational force (slab pull) beneath the adjoining continental crust, creating **subduction** zones

Note that all of this remains a theory because scientists have not yet directly observed the interior of the Earth.

# Plate margins

Key points:

- tectonic hazards (volcanoes, earthquakes, tsunamis) occur at specific points, which are usually associated with tectonic plate margins (Figure 24)

## Key terms

**Constructive (divergent) margins** Where new crust is generated as the plates pull away from each other.

**Destructive (convergent) margins** Where crust is destroyed as one plate dives under another.

**Subduction** The process of one plate sinking beneath another at a convergent plate boundary along a sloping line called a Benioff zone.

## Exam tip

One revision strategy is to sketch and annotate diagrams of different plate margins.

- their distribution is uneven — some areas of the world are at high risk and others are at low risk
- hazards occur at divergent plate margins, convergent plate margins or **conservative plate margins**
- earthquakes also occur where the Indo-Australian plate collides with the Eurasian plate — a continental collision zone
- tectonic events (due to **vulcanicity** and/or **seismicity**) often generate multiple hazards

**Key terms**

**Conservative plate margins** Where crust is neither produced nor destroyed as the plates slide horizontally past each other.

**Vulcanicity** Processes through which gases and molten rock are either extruded on the Earth's surface or intruded into the Earth's crust.

**Seismicity** Processes associated with earthquakes.

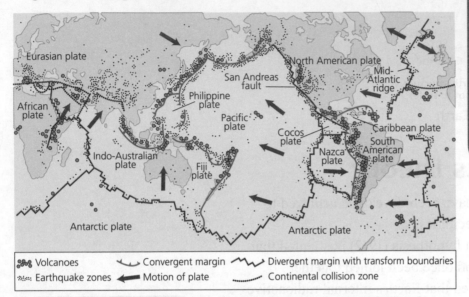

**Figure 24 The global distribution of plate margins, earthquakes and volcanoes**

The combination of tectonic plate motion and plate type yields a range of different plate margin types with varying associated hazards and landforms (Table 10).

**Table 10 Different plate margin settings**

| Margin type | Plate type | Example | Description/hazards |
|---|---|---|---|
| Constructive | Oceanic/oceanic | Mid-Atlantic ridge | Rising convection currents bring magma to the surface, forcing two plates apart and creating ocean ridges; basaltic eruptions; low viscosity of lava; low-magnitude, shallow (<70 km) earthquakes |
| | Continent/continent | East African rift valley/Red Sea | Mantle plume splitting continental plate to create rift valleys; basaltic volcanoes and low-magnitude earthquakes |
| Destructive | Oceanic/oceanic | Aleutian Islands | Subduction of one plate beneath another; frequent earthquakes and an arc of volcanic islands with violent eruptions |
| | Continent/continent | Himalayas | Collision of two continental landmasses (Indo-Australian and Eurasian), creating fold mountain belt; infrequent high-magnitude, shallow earthquakes; volcanoes absent |
| | Oceanic/continent | West coast of South America, Andes mountains | Subduction of oceanic plate beneath continental plate; deep sea trenches and fold mountain range; frequent deep (up to 700 km), high-magnitude earthquakes and violent volcanoes due to high viscosity of lava |
| Conservative | Oceanic/continent | San Andreas Fault | Plates slide past each other along transform faults; frequent shallow, and some high-magnitude, earthquakes; volcanoes absent |

# Magma plumes

Some intra-plate earthquakes and volcanoes do not occur at plate boundaries (e.g. Hawaii and the Galápagos Islands):

- earthquakes can occur in mid-plate settings, usually associated with ancient fault lines being reactivated by tectonic stresses
- volcanoes can occur at **hot spots** — isolated plumes of convecting heat (mantle plumes) rise towards the surface, generating basaltic volcanoes
- a mantle plume is stationary, but the tectonic plate above moves slowly over it
- over millennia, this produces a chain of volcanic islands, with extinct ones most distant from the plume location

> ## Key term
>
> **Hot spot** Area where heat under the Earth's crust is localised — at such points, rising magma can produce volcanoes.

> ## Exam tip
>
> Familiarise yourself with the names of volcanoes associated with each of destructive plate margins and constructive plate margins. Use an atlas.

> ## Do you know?
>
> 1 State two differences between oceanic and continental plates.
> 2 Explain why some geologists suggest that Iceland is a hot spot.
> 3 Why is plate tectonics still just a theory?
> 4 Which type of plate boundary produces the most hazardous volcanoes?
> 5 What is palaeomagnetism and how does it support the idea of sea-floor spreading?

# 3.3 Volcanic hazards

> ## You need to know
>
> - the nature of vulcanicity and its relation to plate tectonics
> - the characteristics of volcanoes — forms, impacts, distribution, magnitude, frequency, regularity and predictability
> - responses to volcanic impacts

# Vulcanicity

Key points:

- volcanoes are built by the accumulation of their own eruptive products: lava, bombs (crusted-over ash deposits) and tephra (airborne ash and dust); gases are also emitted
- a volcano is most commonly a conical mountain built around a vent that connects with reservoirs of molten rock below the surface

- a few volcanoes erupt more or less continuously (e.g. Mauna Loa, Hawaii), but others lie dormant between eruptions
- the type of volcano and volcanic activity depends on the nature of the lava (Table 11):

Table 11 Variations in the type of volcanic activity in relation to types of lava

|  | Basaltic lava | Andesitic lava | Rhyolitic lava |
|---|---|---|---|
| Silica content | 45–50% | 55–60% | 65% |
| Eruption temperature | 1000°C+ | 800°C | 700°C |
| Viscosity and gas content | Very runny, low gas | Sticky, intermediate gas | Very sticky, high gas |
| Volcanic products | Very hot, runny lava Shield volcanoes Low land or plateaux | Sticky lava flows, tephra, ash, gas Composite volcanoes | Pyroclastic flows, gas and volcanic ash Dome volcanoes |
| Eruption interval | Can be almost continuous, as on Hawaii | Decades or centuries | Millennia |
| Tectonic setting and plate margins | Oceanic hot spots and constructive margins | Destructive plate margins: oceanic/continental and oceanic/oceanic | Continental hot spots and continental/continental margins |
| Processes | Dry partial melting of the upper mantle/lower lithosphere. Basaltic magma is generally uncontaminated by water | Wet partial melting of subducting oceanic crust Contaminated by water and other material as magma rises | In situ melting of lower continental crust Most rhyolitic (granitic) magmas cool before they reach the surface |
| Hazardous? | Not really | Very | Very (but rare) |

# Characteristics

## Forms and impacts

Key features:
- lava flows — molten lava is not usually a major threat, but it creates extensive areas of solidified lava
- pyroclastic flows — very hot (800°C), high-velocity (over 200 km/hour) flows of a mixture of gases and tephra, which devastate everything in their path
- ash falls (tephra) — solid material of varying grain size (from fine ash up to volcanic bombs) ejected into the atmosphere. Buildings often collapse under the weight of ash falling on roofs. Air, thick with ash, is very difficult to breathe in and can cause respiratory problems
- nuées ardentes — glowing clouds of hot gas, steam, dust, volcanic ash and larger pyroclasts produced during a violent eruption, which can travel at high velocity

**Exam tip**

A volcano's impact can be judged in terms of its primary and secondary effects, and the environmental, social, economic and political consequences, short and long term. Make sure you include these in your case study of a recent volcanic event.

■ volcanic gases — include carbon dioxide, carbon monoxide, hydrogen sulfide, sulfur dioxide and chlorine. Can be poisonous and contribute to acid rain

Volcanoes also produce **secondary hazards**:

■ mud flows (lahars) — occur when rain mobilises deposits of volcanic ash
■ flooding — caused by the melting of ice caps and glaciers, such as glacial bursts (jökulhlaup in Iceland)

## Other characteristics

Spatial distribution:
■ see Figure 24

Magnitude:
■ measured by the volcanic explosivity index (VEI), based on the volume, duration and column height of ejections
■ can be related back to the type of plate boundary the volcano is located on:
    □ effusive eruptions of basaltic lavas with low VEI (0 to 3) are associated with constructive boundaries or plumes
    □ explosive eruptions of andesitic or rhyolitic lava with high VEI (4 to 7) are associated with destructive boundaries

Frequency:
■ can be determined by previous history if occurred within living memory
■ volcanologists can examine previous deposits, both in the vicinity of the volcano and further afield

Regularity and predictability:
■ see Table 11

# Responses

As volcanic hazards cannot be prevented, the management of, and involve responses to, volcanic events involve:

Preparedness (prediction):
■ study the eruption history of the volcano
■ measure gas emissions, land swelling, groundwater levels
■ measure the shock waves generated by magma travelling upwards

Mitigation (protection):
■ assess the hazard — determine the areas of greatest risk and carry out land-use planning

## Key term

**Secondary hazard** The indirect effects that result from the main (primary) impacts of a hazard.

## Exam tips

■ You have to study *one* case study of a recent volcanic event in detail, but be aware of the management of, and responses to, other volcanic events (although this does not require quite as much detail).
■ You are required to study one recent volcanic event, its impacts and the human responses to it. Possible volcanoes include: Mt St Helens, Mt Nyiragongo, Mt Etna, Montserrat, Mt Merapi and Eyjafjallajökull.
■ Although you are encouraged to keep up to date with events, when undertaking this case study it is recommended that you choose a volcanic event that has run its course and is at least 2–3 years old. In this way all of the requirements can be met.

- dig trenches or carry out explosions to divert the lava
- build barriers to slow down lava flows
- pour water on the lava front to slow it down

Adaptation:

- avoid living in areas at risk
- ensure evacuation routes are available and clear

## Do you know?

**1** Describe the global distribution of volcanoes.

**2** Why did the Eyjafjallajökull eruption have such a great impact on the world?

# 3.4 Seismic hazards

## You need to know

- the nature of seismicity and its relation to plate tectonics
- the characteristics of earthquakes — impacts, distribution, magnitude, frequency, regularity and predictability
- responses to earthquake impacts

# Seismicity

Key points:

- as the Earth's crust is mobile, a slow build-up of stress within the rocks can exist where movement is taking place
- when this stress is suddenly released (when the strain overcomes the elasticity of the rock) parts of the surface experience an intense shaking motion that lasts for a few seconds (an earthquake)
- large amounts of energy are released — much of the energy is transferred vertically to the surface and then moves outwards from the epicentre as seismic waves
- at the moment of fracture rocks may regain their original shape but in a new position
- the depth of focus of an earthquake is significant (Figure 25):
  - □ shallow earthquakes (0–70 km) (75% of all energy released) cause the most damage
  - □ intermediate (70–300 km) and deep (300–700 km) earthquakes have much less effect

## Key terms

**Epicentre** The point on the Earth's surface directly above the focus of an earthquake.

**Focus** The point below the Earth's surface where an earthquake occurs.

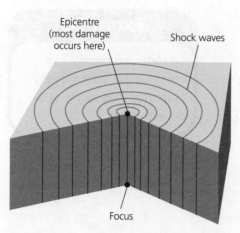

**Figure 25 The focus and epicentre of an earthquake**

Earthquakes generate three types of seismic (shock) wave:

- P-waves (primary waves) are the fastest; they arrive first and cause least damage
- S-waves (secondary waves) arrive next and shake the ground violently
- L-waves (Love waves) arrive last as they travel only across the surface; they have a large amplitude and cause significant damage, including fracturing of the ground

# Characteristics

## Impacts

Secondary hazards:

- large landslides — especially in mountainous areas such as Nepal
- liquefaction — where the ground consists of loose sediments of silts, sands and gravels that are waterlogged. Earthquakes compact the sediments and force the water to the surface, undermining buildings and roads

Tsunamis:

- most tsunamis are generated by submarine earthquakes at subduction zones
- the seabed is displaced vertically (up or down)
- this motion displaces a large volume of water in the ocean column, which then moves outwards from the point of displacement
- when they are out at sea they have a very long wavelength, often in excess of 100 km
- they are very short in amplitude, at around 1 m in height, and are barely noticeable
- they travel very quickly, often at speeds of up to 700 km h$^{-1}$

> ### Exam tip
>
> An earthquake's impact can be judged in terms of its primary and secondary effects, and the environmental, social, economic and political consequences, short and long term. Make sure you include these in your case study of a recent seismic event.

- when they reach land they rapidly increase in height, up to over 25 m in some cases
- they are often preceded by a localised drop in sea level (drawback) as water is drawn back and up by the approaching tsunami
- they hit a coastline as a series of waves (a wave-train), more akin to a flood

> **Exam tip**
>
> Despite commonly being called 'tidal waves', tsunamis have nothing to do with tides.

## Other characteristics

Spatial distribution:
- see Figure 24

Magnitude:
- energy release is measured by the logarithmic moment magnitude scale (MMS), a modification of the Richter scale
- damaging effects are measured by the Mercalli scale (measures intensity of shaking)

Frequency:
- can be determined by previous history if occurred within living memory
- seismologists can examine previous faults and movements in the field

Regularity and predictability:
- geologists use 'gap theory' — where there has been a 'gap' in time since a previous event, then the likelihood of an event in the future increases
- the notion of the 'next big one' is very unpredictable, however

## Responses

Preparedness (prediction):
- the study of groundwater levels; the release of radon gas and animal behaviour
- the monitoring of fault lines and local magnetic fields
- the study of fault lines to look for 'seismic gaps' where the next earthquake may occur

Prevention:
- keeping the plates sliding past each other rather than 'sticking' and then releasing; suggestions include using water and/or oil

Mitigation (protection):

■ see Figure 26

## Key term

**Earthquake kits** Boxes of essential household supplies (water, food, battery-powered radio, blankets) kept in a safe place at home to be used in the days following an earthquake.

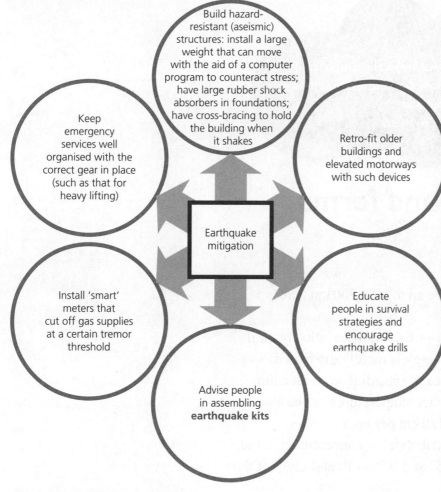

**Figure 26 Earthquake mitigation**

Adaptation:

■ planning land use to avoid buildings being constructed in high-risk areas (e.g. low-lying coasts in tsunami-prone areas)
■ avoiding areas where liquefaction is likely

## Do you know?

**1** Why did the Tohoku earthquake and tsunami have such a great impact on the world?

**2** Can earthquakes be predicted?

**3** What is aseismic design?

## Exam tips

■ You have to study *one* case study of a recent seismic event in detail, but be aware of the management of, and responses to, other earthquakes (although this does not require quite as much detail).

■ You are required to study one recent seismic event, its impacts and the human responses to it. Possible earthquakes include: Northridge (Los Angeles), Gujarat (India), Banda Aceh (Indonesia), L'Aquila (Italy), Tohoku (Japan), Haiti, Christchurch (New Zealand), Ghorka (Nepal) and Amatrice (Italy).

■ Although you are encouraged to keep up to date with events, when undertaking this case study it is recommended that you choose one that has run its course and is at least 2–3 years old. In this way all of the requirements can be met.

# 3.5 Storm hazards

## You need to know

- the nature, causes and forms of tropical storm hazards
- the characteristics of storms — distribution, magnitude, frequency, regularity and predictability
- storm impacts and responses to them

# Nature, causes and forms

## Nature and causes

Nature of storms:

- systems of intense low pressure up to 600–700 km across (Figure 27)
- formed over tropical sea areas — they move erratically until they reach land, where their energy is rapidly dissipated
- at their centre they have an area of subsiding air with calm conditions, clear skies and higher temperatures — the eye
- have wind speeds exceeding 120 km per hour
- predictable in their spatial distribution — concentrated in the tropics, specifically between 5° and 20° north and south of the equator
- once generated they tend to move westwards initially, before switching to a northeastward direction as they move further away from the equator, and 'die'

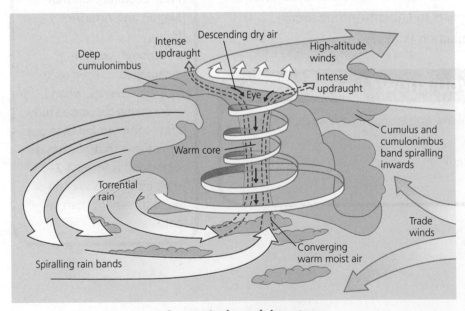

**Figure 27 The structure of a tropical revolving storm**

Causes of storms:
- begin with an area of low pressure in the tropics, into which warm air is drawn in a spiralling manner
- small-scale disturbances enlarge into storms with rotating wind systems that become more intense and rapid

Factors determining storm growth:
- an oceanic location where sea temperatures are over 27°C
- a location at least 5° north or south of the equator, so that the effect of the Coriolis force can bring about the maximum rotation of the air
- rapidly rising moist air (from the warm sea) cools and condenses, releasing latent heat energy that then fuels the storm
- low-level convergence of air occurs in the lower part of the system, but this is then matched by intense upper atmosphere divergence of air, together creating a strong updraught of air
- they fade and 'die' over land as the energy source is removed

## Forms

Hazards:
- high winds — cause structural damage and collapse of buildings and bridges
- storm surges result from the piling up of water by wind-driven waves and the ocean rising up under reduced atmospheric pressure
- coastal flooding can extend inland if the area near the coast is flat and unprotected
- river flooding caused by heavy rainfall (over 100 mm a day)
- landslides in areas of high relief

## Other characteristics

Spatial distribution:
- found in the Caribbean Sea, Bay of Bengal, South China Sea, east of Japan, northern Australia

Magnitude:
- measured by the Saffir–Simpson scale, with five categories
- category 4 storms have winds over 200 km h$^{-1}$; category 5 over 250 km h$^{-1}$

Frequency:
- each year there are about 80 to 100 storms, though there is no pattern to their severity
- some suggest that their frequency is increasing with climate change, although the evidence for this is unclear

### Key term

**Coriolis force (CF)** The effect of the Earth's rotation on air flow. In the northern hemisphere the CF causes a deflection in the movement of air to the right, whereas in the southern hemisphere it is to the left.

### Exam tip

The forms of hazard associated with a tropical storm will vary according to the individual storm studied. When examining your two chosen case studies of recent tropical storms, make sure you note the forms of hazard for those storms.

### Exam tip

A storm's impact can be judged in terms of its primary and secondary effects, and the environmental, social, economic and political consequences, short and long term. Make sure you include these in your two case studies of recent storm events.

Regularity and predictability:

■ hurricanes in the Caribbean are predictable in their timing and frequency — usually between September and November

# Responses

Preparedness:

■ predicting the origin and tracks of storms depends on the quality of monitoring and warning systems. The USA maintains round-the-clock surveillance of hurricanes using aircraft
■ schoolchildren in Florida practise hurricane drills — part of an awareness programme (Project Safeside)
■ better computer modelling enables forecasters to use sophisticated methods to predict changes in wind speed, humidity, temperature and cloud cover, e.g. the NOAA

Mitigation:

■ directed at ways of reducing the storm's energy while it is still over the ocean
■ one attempt has been to 'seed' the storm using silver iodide outside of the eye-wall clouds to produce rainfall, so releasing latent heat that would otherwise sustain the high wind speeds

Adaptation and prevention:

■ see Figure 28

### Key term

NOAA National Oceanic and Atmospheric Administration.

### Exam tips

■ You are required to study *two* recent tropical storms in contrasting areas — their impacts and the human responses to them. Possible storms include: Cyclone Nargis (Myanmar), Hurricane Katrina (USA), Typhoon Haiyan (the Philippines), Hurricane Sandy and Hurricane Matthew (both the Caribbean and USA).
■ Although you are encouraged to keep up to date with events, when undertaking these case studies it is recommended that you choose two that have run their course and are at least 2–3 years old. In this way all of the requirements can be met.

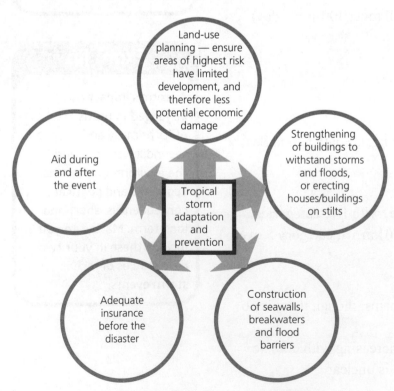

**Figure 28 Tropical storm adaptation and prevention**

**Do you know?**

1 What are the local names for tropical storms around the world?

2 Why is it essential that storm warnings are accurate?

3 How does the impact of a tropical storm depend on political and economic factors?

4 In the context of tropical storms, what is meant by 'cry wolf syndrome'?

# 3.6 Fires in nature and multi-hazardous areas

**You need to know**

■ the nature of, and conditions favouring and causing, wildfires

■ the impacts of wildfires and responses to them

■ the concept of multi-hazardous environments with reference to an example beyond the UK

## Nature and causes

Key points:

■ commonly known as bushfires (Australia) or brush fires (USA/Canada)

■ many occur naturally (10%) — the result of lightning strikes

■ most are human-induced fires that have gone out of control (cannot be classed as managed) or have been started deliberately with malicious intent or through carelessness

■ the spread of a wildfire depends on the types of plants involved, the topography, the strength and direction of the wind, and the relative humidity of the air in the region

■ some travel close to the ground; others spread via the canopies of tall trees

■ associated with areas with semi-arid climates where there is enough rainfall for vegetation to grow and provide a 'fuel', yet with a dry season to promote conditions for ignition

■ traditionally wildfires have not been associated with areas of tropical rainforest because of the high humidity and all-year-round rainfall

■ the burning of the rainforest (e.g. in Indonesia) for clearance and by logging companies, however, combined with the drying effect of El Niño events, has revised this view

■ such fires are smoky events with toxic hazes, combined with other forms of air pollution, hanging over cities such as Kuala Lumpur and Singapore

# Impacts and responses

## Impacts

General impacts:

■ loss of timber, livestock and crops
■ loss of plant species and the creation of areas dominated by fire-resistant scrub
■ damage to soil structure and nutrient content over a wide area
■ loss of wildlife, a concern when rare or endemic species are involved
■ if fire is extensive, the loss of vegetation can lead to an increased risk of flooding
■ temporary evacuation and emergency aid for the areas affected
■ property loss is increasing as a result of settlement expansion into at-risk areas
■ release of toxic gases and particulate pollution
■ loss of life is usually quite low; firefighters are at greatest risk
■ huge costs for emergency services and large numbers of people involved in controlling the outbreak

## Responses

Preparedness:

■ use of technology — to warn areas at risk. Aircraft and satellites used to carry out infrared sensing to measure ground surface temperatures and check for signs of eco-stress from desiccation
■ community preparedness — using fire towers; training people to act as auxiliary firefighters, to organise evacuation and coordinate emergency firefighting

Mitigation:

■ in Australia, Greece and Spain, the main approach to fire management is to extinguish fires as they occur, especially in populated areas or near to high-value timber reserves
■ also involves reducing or eliminating the fuel supplies from the potential path of the fire — by controlled burning. This practice is not only controversial, but also risky

> ### Exam tips
>
> ■ When examining your chosen case study of a recent wildfire event, make sure you note the characteristic impacts of that event.
> ■ The impact of a wildfire can be judged in terms of its primary and secondary effects, and the environmental, social, economic and political consequences, short and long-term.

Prevention:

■ education concerning home safety in high-risk areas
■ reduction of supplies of fuel, wood stores stacked correctly, and adequate water hoses and ladders available
■ householders are advised to remove dead leaves from gutters
■ school education concentrates on ensuring young people understand the dangers of arson and casual cigarette use, and the need to adhere to barbeque laws

Adaptation:

■ see Figure 29

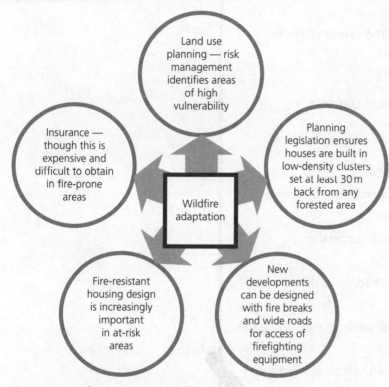

**Figure 29** Adaptation to wildfires

# Multi-hazardous environments

You are required to study one **multi-hazardous environment** beyond the UK to illustrate and analyse the nature of the hazards and the social, economic and environmental risks presented, and how human qualities and responses such as resilience, adaptation, mitigation and management contribute to the area's continuing human occupation.

Possible areas include Los Angeles, Tokyo, Bangkok, Manila and Kolkata.

## Do you know?

1 Why are managed fires used in some areas?
2 Describe the spatial distribution of wildfires.
3 Some plants are pyrophytic. Explain what this means and give some examples.
4 Why do some of the megacities in both the developed and the developing world have the potential to be multi-hazardous areas?

## Exam tip

Although it may be an obvious point, it is advised that your chosen multi-hazardous environment is affected by at least one of a volcano, an earthquake, a tropical storm or a wildlfire.

## End of section 3 questions

1 In the context of tropical storms, outline the causes of storm surges.

2 Explain the causes of tsunamis.

3 Explain the development of hazards found at convergent (destructive) plate margins.

4 Outline the causes of tropical storms.

5 Describe how earthquakes can be measured.

6 Evaluate the effectiveness of the hazard management cycle in assisting with the planning for wildfire events.

7 Assess the importance of governance in the successful management of tectonic disasters.

8 Evaluate the management of, and response to, one seismic event you have studied.

9 Assess the importance of factors of globalisation in supporting the response to major seismic hazards.

10 Discuss how well the hazards associated with tropical storms can be managed.

# 4 Global systems and global governance

## 4.1 Globalisation

### You need to know

- the dimensions of globalisation
- global patterns of production, distribution and consumption
- the factors that influence globalisation

## Dimensions

Key points:

- **globalisation** has caused the world to become more interconnected (Figure 30)
- such interconnections have been driven by flows of:
  - □ goods — products, commodities and services
  - □ capital — money between people, banks, businesses and governments
  - □ people — migrants (economic and refugees) and tourists
  - □ ideas — information, both political and cultural

### Key term

**Globalisation** The growing economic interdependence of countries worldwide through the increasing volume and variety of cross-border transactions in goods and services, freer international capital flows, and more rapid and widespread diffusion of technology.

### Exam tip

Note that the concept of globalisation is not new — the process began centuries ago. The use of the internet has meant that the current level of change is fast, however.

Figure 30 Factors and dimensions of globalisation

Consequently, globalisation is said to have a number of dimensions (Table 12).

**Table 12 Dimensions of globalisation**

| | |
|---|---|
| Economic globalisation — the growth of global transnational corporations (TNCs), which have a global presence and global brand image; also involves the spread of investment around the globe and rapid growth in world trade | Political globalisation — the dominance of western democracies in political and economic decision making; also spreads the view that democratic, consumerist societies are the most 'successful' |
| Social globalisation — occurs as migration and tourism increase; populations are becoming ever more fluid and mixed | Cultural globalisation — involves people increasingly eating similar food, wearing similar clothes, listening to similar music and sharing similar values, many of which are 'western' in origin, i.e. from North America and Europe |
| Environmental globalisation — an awareness of global environmental concerns, such as climate change and threats to Antarctica, and of the need to address them | |

# Global marketing

When a company decides to embark on **global marketing**:
- it views the world as one market
- its ultimate goal is to sell the same thing, the same way, everywhere — creating a global brand
- it modifies products to fit the various regional marketplaces (glocalisation)
- a common factor seems to align a product with markets in the USA and the Far East

# Global patterns

## Production

Globalisation has created an international division of labour:
- occupations that are highly skilled, highly paid and involve R&D, decision making and managerial roles — these are largely concentrated in developed (or high-income) countries
- occupations that are unskilled and poorly paid assembly roles — these tend to be located in the emerging economies, often developing (or low-income) countries, based on their lower labour costs

**Key terms**

**TNCs** Transnational corporations — companies that operate in more than one country (also known as multinationals).

**Global marketing** Marketing on a worldwide scale, taking commercial advantage of global operational differences, similarities and opportunities in order to meet global objectives.

**Glocalisation** Changing the design of products to meet local tastes or laws. It is an increasingly common strategy used by TNCs in an attempt to conquer new markets.

**R&D** Research and development.

**Exam tip**

What products have you seen on your holidays that are the same or similar to those in your country?

This division has arisen from:

■ many countries that were classified as developing have become NICs

■ FDI by TNCs towards those NICs, which enabled manufacturing at a competitive price — a movement called the 'global shift'

■ the transfer of technology, which enabled NICs to increase their productivity without raising their wages to the levels of developed countries

Main outcomes:

■ by 2015 over 50% of manufacturing jobs were located in the NICs, and over 60% of exports from such countries to the developed world were manufactured goods

■ deindustrialisation — not entirely due to the 'global shift', as other factors such as outmoded production methods, long-established products entering the end of their life cycles and poor management also all contributed

## Services (distribution)

Key points:

■ the provision of services has become increasingly detached from the production of goods

■ the 1990s saw the emergence of a growing number of TNC service conglomerates, particularly in banking and other financial services and in advertising

■ the movement of capital around the world has speeded up, such that many service industries are owned by TNCs purely for their financial gain, e.g. private equity firms and venture capitalists

■ the decentralisation of low-level services from the developed to the developing world, e.g. call centre operations in India and South Africa, again due to lower labour costs

## Consumption

Key points:

■ as NICs develop, their populations are becoming more affluent and demanding similar consumer products to the developed world

■ this may mean trade becomes focused towards east Asia, and intra-Asian trade may increase

■ consumption of financial services may increase in Asia, leading to western TNCs expanding there

> ### Key terms
>
> **NICs** Newly industrialised countries, such as China, India, South Korea and Indonesia.
>
> **FDI** Foreign direct investment — money invested into a country by TNCs or other national governments.
>
> **Transfer of technology** The movement of ideas and technology from one region or country to another.
>
> **Deindustrialisation** An absolute, or relative, decline in the importance of manufacturing in the industrial economy of a country and a fall in the contribution of manufacturing to GDP.

# Factors

## The development of technologies

Technological developments that have taken place to create a 'shrinking world' to facilitate globalisation:

- the development of containers — intermodal metal boxes that can be carried by ship, lorry or train
- the growth of the logistics industry to manage the movement of containers
- developments in ICT and mobile technology that enable **time/space compression** — satellites, fast broadband, networks of fibre-optic cables, smartphones
- the use of these technologies by businesses to keep in touch with all elements of production, supply and sales, and to transfer money and investments
- the use of these technologies by individuals for social networking, banking, shopping and leisure activities
- the use of these technologies has led to a range of security-based systems and industries
- security for container transport and air transport, and for financial transactions and cyber-security, are major issues for the globalised world

## Management and information systems

TNCs have modified their management and information systems:

- **Global production networks (GPNs)** — large corporations (ranging from Dell to Tesco) have established multiple subcontracting partnerships while building their global businesses, often coordinated by a **hub company**
- **just-in-time production** — requires a very efficient ordering system and reliability of delivery. Production is said to be 'pulled through' rather than 'pushed through'
- **'zero defect'** — TNCs have very strong links with their suppliers that are monitored rigorously
- the management of all of the above can be done remotely using ICT

## Trade agreements

Most governments actively seek global connections in the belief that trade promotes economic development and wealth by:

- joining **free trade blocs** such as the European Union (EU) and the Association of Southeast Asian Nations (ASEAN),

---

### Key terms

**Shrinking world** The idea that the world feels smaller over time, because places are closer in terms of travel or contact time.

**Time/space compression** The idea that the cost of communicating over distance has fallen rapidly.

**Global production network (GPN)** A system whereby a TNC manages a series of suppliers and subcontracted partnerships while building its global business.

**Hub company** A company that orchestrates production on a global scale.

**Just-in-time (JIT) production** A management system that is designed to minimise the costs of holding stocks of raw materials and components by carefully planned scheduling and flow of resources through the production process.

**Free trade bloc** An agreement between a group of countries to remove all barriers to trade, such as import/export taxes, tariffs and quotas.

---

### Exam tip

Free trade and 'fair trade' are often confused. Make sure you understand the difference.

---

which make trade barrier-free between member states and in the case of the EU allows free movement of people between countries

■ opening up their markets to competition and enabling **free market liberalisation**

## Do you know?

1 Outline some historical aspects of the 'shrinking world'.
2 Give some examples of global marketing.
3 Where are the major financial services located in the world today?
4 Explain how a hub company operates.

# 4.2 Global systems

## You need to know

■ the form and nature of interdependence in the contemporary world
■ issues associated with interdependence (unequal flows of people, money, ideas and technology within global systems)
■ issues associated with interdependence (unequal power relations and geopolitical events)

## Interdependence

Key features:

■ economic **interdependence** through trade can be exemplified by a developed country exporting manufactured goods to a developing country, and importing raw materials in return

■ socio-economic interdependence occurs where economic migrants provide labour in a country and on their return bring back newly acquired skills, ideas and values to their home country

■ increasingly, countries are becoming interdependent in their effects on the global environment and in their political relationships

■ **global governance** is struggling to keep up with the pace and extent of economic globalisation, capital and trade flows, illegal and legal migration of people, and technological change

## Key terms

**Interdependence** The mutual dependence of two or more countries, in which there is a reciprocal relationship.

**Global governance** The emergence of norms, rules, laws and institutions that have regulated and reproduced the trade-orientated global systems, as well as other global systems (such as those involving patterns of population migration).

# Inequality of flows

Unequal flows of people, money, ideas and technology within global systems can:

- sometimes act to promote stability, growth and development
- cause inequalities, conflicts and injustices for people and places

## Stability, growth and development

Stability:

- trade contributes to international peace and stability, especially if countries trade under the same rules, such as those of the WTO
- trade encourages states to cooperate — multilateral and bilateral trade agreements contribute to economic and political stability
- some bilateral agreements extend beyond trade and may lead to cooperation and assistance in dealing with political issues, such as strengthening democratic processes and human rights (e.g. combating child labour)
- all of the above create a more stable environment for foreign investors

Economic growth:

- trade in merchandise and commercial services stimulates production and contributes to GDP growth and to further investment, including FDI
- employment opportunities are created, incomes are raised and in some developing countries poverty levels can be significantly reduced
- the economic multiplier can be enhanced by international trade
- foreign exchange (monetary flow) generated by trade can stimulate further domestic and foreign investment

Development:

- removal of tariffs and other obstacles to trade helps generate foreign exchange, which can be invested to reduce internal inequalities in poverty, health, education, infrastructure and transport
- the corporate social responsibility of TNCs can be of economic and social benefit to employees and communities in areas of production
- membership of regional trading blocs and political unions can help socio-economic development within member states
- migration of highly skilled workers (scientists, engineers) can be innovative in circulating ideas and information on technology development between countries

## Key terms

WTO World Trade Organization.

Economic multiplier An initial investment in an economic activity in an area that has beneficial knock-on effects elsewhere in the area's economy.

Corporate social responsibility A TNC's commitment to assess and take responsibility for its social and environmental impact — including its ethical behaviour towards the quality of life of its workforce, their families, and local communities.

# Inequality, conflict and injustice

Inequality:

- Many developing countries have limited access to global markets — this widens the development gap
- skilled workers, especially men, tend to benefit most from employment opportunities created by trade, whereas many unskilled workers and women are held back by limited education opportunities and remain unemployed and unable to contribute to the workforce
- in developing countries internal inequalities are exacerbated by trade activity, often spatially concentrated in ports, where most commercial activity is located
- leads to internal migrant flows and widening inequalities within the country

Conflict:

- disputes can arise over tariffs, prices of commodities and changes to trade agreements
- border and customs authorities can be subject to corruption and breaches of security
- port development, mining and deforestation create environmental conflicts

Injustice:

- displacement of communities takes place due to land-grabbing by investments in mining or agriculture
- attempts to secure cheap labour can lead to child labour and forms of modern slavery
- unfair trade rules can adversely affect businesses, such as those of small-scale farmers or fishermen

> ## Key term
>
> **Development gap** The difference in prosperity and wellbeing between rich and poor countries — measured by GDP per capita.

> ## Exam tip
>
> Questions on this area are likely to ask for evaluation or assessment. Make sure your answer is discursive and has a clear conclusion.

# Inequality of power relations

Unequal power relations mean that:

- some states are able to drive global systems to their own advantage and to directly influence geopolitical events
- others are able only to respond or resist in a more constrained way

## States that drive global systems

This can be illustrated with reference to a developed economy, such as China or the USA, and how it has a strong geopolitical influence and drives the global system to its own advantage as a result of some or all of the factors in Table 13.

**Table 13** Factors that help a state drive global systems, and the challenges it may face

| Factor | Features |
|---|---|
| The components of its international trade | ▪ its patterns of trade, trade partners and trade agreements<br>▪ its membership of trading blocs |
| Its advantages for international trade | ▪ investment in domestic transport and communications infrastructure<br>▪ industrial productivity<br>▪ outward FDI<br>▪ ability to exploit its own natural resources<br>▪ political strength in negotiating trade agreements<br>▪ levels of skill and education in the workforce |
| Opportunities that international trade creates for the country | ▪ employment in a wide range of industrial sectors<br>▪ stimulation of the economic multiplier effect<br>▪ development of positive political and cultural relationships with its trade partners, including stewardship of the environment<br>▪ ability to integrate other countries, rich and poor, into its supply chains |
| Challenges that arise as a result of its influence | ▪ pollution issues and land-use conflicts, e.g. resulting from port development<br>▪ trade disputes over 'price dumping'<br>▪ managing a trade surplus or deficit<br>▪ managing migration across its borders |

# States that respond to global systems

This can be illustrated with reference to a developing economy, such as a sub-Saharan country, and how it has a limited influence on, and can only respond to, the global system as a result of some of the factors given in Table 14.

> **Exam tip**
>
> When discussing this topic, try to write about a range of economic, social and environmental aspects.

**Table 14** Factors that influence a state having to respond to global systems

| Factor | Features |
|---|---|
| The components of its international trade | ▪ its patterns of trade, trade partners and trade agreements<br>▪ its membership of trading blocs |
| Its limited access to global markets | ▪ its limited ability to exploit, transport, market and export its primary products<br>▪ its inability to cope with economic shock, such as changes in global demand and prices for primary products<br>▪ its vulnerability to natural hazards, and the effects of political shock, such as conflict, on its economy |
| Opportunities or otherwise brought by international trade | ▪ economic development and diversification of industry<br>▪ need for investment in key infrastructure<br>▪ socio-economic development through investment in health and education<br>▪ human rights issues, crime and conflict<br>▪ interdependence with the EU and USA, and similar |
| Challenges that remain | ▪ achieving political stability and democracy<br>▪ removing barriers that prevent integration into global systems, such as illegal practices or limited investment in transport infrastructure<br>▪ managing environmental problems arising from mining and deforestation operations<br>▪ reducing socio-economic inequalities |

**Exam tip**

Be aware of unequal power relations within the global system — identify some of the world's most powerful and least powerful countries.

# 4.3 International trade and access to markets

**You need to know**

■ the main features and trends of international trade and investment

■ the trading relationships and patterns between developed economies, emerging economies and less developed economies

■ how differential access to markets is associated with economic development and societal wellbeing

■ the nature, role and other characteristics of transnational corporations

■ world trade in one food commodity or manufactured good

■ how globalisation impacts the wider world

## International trade and investment

### Trade

The global pattern of international **trade** is uneven and complex:

■ it is dominated by developed countries (USA and Germany) and the faster-growing emerging countries (China and India)

■ low-income developing countries (e.g. in sub-Saharan Africa) have limited access to international markets and as a result have relatively weak terms of trade

■ countries that export manufactured goods earn more from trade than those that export raw materials — manufactured goods have a higher value than raw materials

■ changes are taking place in global patterns of trade:

☐ emerging countries now trade more with each other than with developed countries

☐ developed countries are exporting less to each other and more to the emerging countries

**Exam tip**

This section provides just the main trends. Visit **www.wto.org** for up-to-date statistics.

**Key term**

Trade  The movement of goods and services from producers to consumers.

- commercial services can also be traded:
  - ☐ the largest exporters are the EU, the USA and, increasingly, China and India
  - ☐ there are very low provisions for the countries of sub-Saharan Africa

## Investment

Key features:
- FDI by transnational corporations (TNCs) in foreign countries is a key element of the global system — most currently emanates from developed countries
- the emerging economies, particularly China and the Gulf states, are increasing their levels of FDI
- much Chinese investment is in developing countries, such as those in sub-Saharan Africa, aimed at protecting their supply lines of raw materials

## Trading relationships

See Table 15.

**Table 15** International organisations that promote trade and investment

| Organisation | Commentary |
|---|---|
| World Trade Organization (WTO) | ■ the WTO works to reduce trade barriers and create free trade — previously known as GATT<br>■ a series of global agreements has reduced trade barriers and increased free trade<br>■ the latest round of talks began in Doha in 2001 but has not been agreed yet |
| International Monetary Fund (IMF) | ■ since 1945 the IMF has worked to promote global economic and financial stability, and to encourage more open economies<br>■ part of this involves encouraging developing countries to accept FDI and open up their economies to free trade<br>■ the IMF has been criticised for promoting a 'western' model of economic development that works in the interests of developed countries and their TNCs |
| World Bank | ■ the World Bank's role has been to lend money to the developing world to fund economic development and reduce poverty<br>■ it has helped developing countries develop deeper ties with the global economy, but has been criticised for having policies that put economic development before social development |
| North American Free Trade Area (NAFTA) | ■ NAFTA's members consist of the USA, Canada and Mexico<br>■ its main impact has been to create the maquiladora in Mexico; this is off-shoring, based entirely on numerous low-cost labour forces in north Mexico |
| Association of Southeast Asian Nations (ASEAN) | ■ a political and economic organisation of ten Southeast Asian countries — Indonesia, Malaysia, the Philippines, Singapore, Thailand, Brunei, Cambodia, Laos, Myanmar and Vietnam<br>■ its aims include accelerating economic growth and social progress, and promoting regional peace and stability |

# Differential access to markets

Key points:

- differential access to markets impacts on the economic and societal wellbeing of nations
- the global pattern of trade is one of great inequality, with many developing countries, or regions of countries, having limited access to global markets

A number of strategies seek to deal with these issues:

- Special and Differential Trading agreements (SDTs) between some developing nations and some developed countries — sanctioned by the WTO. These encourage export diversification to reduce over-dependency on single exports
- Free-Trade Areas (FTAs), e.g. the region of north Mexico (maquiladora) — in recent years this has come under fire from President Trump for undercutting US industries
- Special economic zones (SEZs):
  - ☐ tariff- and quota-free, allowing manufactured goods to be exported at no cost
  - ☐ infrastructure such as port facilities, roads, power and water connections are supplied by the government, providing a subsidy for investors and lowering their costs
  - ☐ taxes are very low, and often there is a tax-free period of up to 10 years after a business invests
  - ☐ unions are usually banned, so workers cannot strike or complain
  - ☐ environmental regulations are usually limited

# Transnational corporations

Key features:

- dominate international trade — TNCs are large companies that are important drivers of the global trade system
- occur in a wide range of industries — resource extraction (oil, metal ores), manufacturing (electronics, food, cars) and services (banking, supermarkets, hotels)
- responsible for developing global supply chains through investments in factories and businesses in emerging countries, such as India and China
- make further TNC investments in sub-Saharan Africa, Southeast Asia and Latin America

## Key terms

**Maquiladora**
Manufacturing industries operating in a Mexican free trade zone close to the USA/Mexico border, where factories import material and equipment on a duty-free and tariff-free basis for assembly, processing or manufacturing. The products are then re-exported back to the USA and Canada.

**Off-shoring** The manufacture or assembly of a product in a developing country using components produced in a developed country.

## Exam tip

Be prepared to evaluate the costs and benefits of SEZs and FTAs.

## Spatial organisation

Key features:

- most have headquarters and R&D in developed countries (Europe and the USA)
- manufacturing is usually based in areas of low labour costs — Southeast Asia and eastern Europe

## Production

Key features:

- most take advantage of **outsourcing** their production — some subcontracting arrangements can be highly complex (global production networks, see page 66)
- many TNCs also outsource their back-office and other services

see page 66

> ### Key term
>
> **Outsourcing** A TNC sub-contracts an 'overseas' company to produce goods or services on its behalf.

## Linkages

There are two different types of 'operation':

- overseas branch plant operations — production facilities owned by the parent company. An example would be Ford, a US car company with its headquarters in the USA, but with branch plants in the UK, Belgium and Mexico
- global production networks (GPNs)

## Trading and marketing patterns

There are two broad types of pattern:

- vertical integration — a supply chain of a company is owned entirely by that company, from raw material to finished product
- horizontal integration — a company diversifies its operations by expansion, merger or takeover to give a broader capability at the same stage of production

## A specified TNC

You are required to study a specified TNC, including its impact on those countries in which it operates. Table 16 provides general impacts.

> ### Exam tip
>
> Choose a company that you can research easily and that will interest you. Examples include Apple, Ford and Tata.

**Table 16 Benefits and costs of TNCs**

| | For the host country | For the TNC | For the country of origin |
|---|---|---|---|
| **Benefits** | Generate jobs and income <br><br> Bring new technology <br><br> Develop new skills <br><br> Create a multiplier effect | Lower costs because of cheaper land and lower wages <br><br> Access to new resources and markets <br><br> Fewer controls on environmental matters | Cheaper goods for sale <br><br> Can specialise in aspects of production, such as R&D |
| **Costs** | Poor working conditions — sweatshops <br><br> Exploitation of resources <br><br> Negative impacts on the environment and culture <br><br> Repatriation of profits <br><br> Possibility of corruption | Ethical issues that may impact on sales <br><br> Reputational damage | Loss of manufacturing jobs <br><br> Deindustrialisation <br><br> Structural unemployment <br><br> Political discontent |

# World trade in a commodity or product

You are required to study the world trade in at least one food commodity or one manufactured product, for example cars (Figure 31).

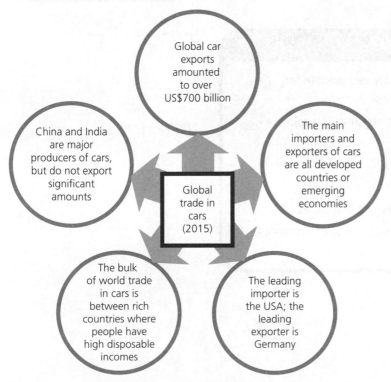

**Figure 31 Global trade in cars (2015)**

# Globalisation critique

Key points:

- as globalisation continues apace and nations desire to raise the living standards of their people, it is apparent that economic development cannot rise smoothly alongside social, cultural and environmental development
- some have argued that globalisation helps integrate the world, thereby maintaining peace and stability
- it is also true that the richest nations, as well as individuals, want to maintain their differential, even if their desire to do so impacts negatively on others
- globalisation is driven by decision makers of world governance — supranational, national or individual
- others believe that globalisation has created greater inequality, injustice, conflict and environmental degradation
- they point to issues such as deforestation, water pollution, climate change and biodiversity loss
- consequently, various anti-globalisation movements have arisen
- some places in the world are still 'switched off' from globalisation due to political isolation (North Korea), physical isolation (Bhutan and Chinese Tibet) and economic isolation (the Sahel of Africa)

## Exam tip

It is important that you develop a view on these issues, and consider how others may see them differently from you.

## Do you know?

1 Which global organisation has been mainly responsible for removing trade barriers between countries?

2 Assemble three arguments for and three arguments against a country staying in the EU.

3 When did China first open itself up for FDI?

4 Outline the impact of SEZs in China.

5 Name some examples of TNCs in a variety of industries.

6 Give one example of vertical organisation and one of horizontal organisation in a TNC.

# 4.4 Global governance

## You need to know

■ how norms, laws and institutions operate within global systems

■ issues associated with global governance, with specific reference to the United Nations

## Norms, laws and institutions

Key points:

■ global governance has shaped relationships between states and non-state organisations (e.g. the United Nations (UN), TNCs and NGOs) (Figure 32)

■ many of the laws and institutions have been responsible for positive changes in the way in which global geopolitics operate

■ UN-sponsored agreements on **human rights** and genocide, coupled with international law, were crucial in creating the post-1945 international system

■ the UN treats nation states as equal partners under the auspices of its Charter

■ global governance has dealt with matters concerning trade, security, nuclear proliferation, human rights, sovereignty and territorial integrity, the atmosphere, laws of the sea and the protection of animals

■ some of these are referred to as the 'global commons' — the Earth's resources that are, in theory, shared by all (see page 79)

### Key terms

**NGO** Non-governmental organisation.

**Human rights** Moral principles or norms that describe certain standards of human behaviour, and are protected as legal rights in international law.

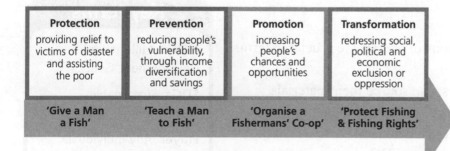

| Protection | Prevention | Promotion | Transformation |
|---|---|---|---|
| providing relief to victims of disaster and assisting the poor | reducing people's vulnerability, through income diversification and savings | increasing people's chances and opportunities | redressing social, political and economic exclusion or oppression |
| 'Give a Man a Fish' | 'Teach a Man to Fish' | 'Organise a Fishermans' Co-op' | 'Protect Fishing & Fishing Rights' |

**Figure 32 Different roles of NGOs**

# Issues

There have been two significant issues associated with global governance:

- how agencies, including the UN in the post-1945 era, have worked to promote growth and stability, and yet may also have created and exacerbated inequalities and injustices
- how the interactions between the local, regional, national, international and global scales have become fundamental to understanding the role of global governance

# The United Nations

Key features:

- includes 193 countries as its member states, and its main headquarters is located in New York, USA
- believes that only through international cooperation can we meet the challenges facing the world
- operates through applying the principles of its Charter; its main authority in maintaining international peace and security is the Security Council
- does not maintain its own military — its peacekeeping forces are supplied by member states
- aims to protect human rights by adopting the Universal Declaration of Human Rights as a standard
- provides food, drinking water, shelter and other humanitarian services to people displaced by famine, war and natural disaster
- in both 2000 and 2015, the UN set targets (the MDGs and SDGs, respectively) regarding tackling poverty, child and maternal mortality, diseases and environmental concerns

## Exam tip

Be able to examine the success or otherwise of two specific MDG/SDG targets.

# Interactions

The interdependence and interactions of players are crucial in terms of governance. Players can include:

- TNCs — can form cooperatives; can source their materials and products ethically; can enforce codes of conduct for their producers; or they can deliberately do none of these
- national governments — can regulate TNCs
- supranational bodies (EU, WTO) — regulate trade
- workers — can form trade unions to defend their rights; in extreme cases they can cite solidarity internationally
- consumers — can ask moral questions about the origin of food and other products, and they can reject exploitative goods

## Key terms

**MDGs** Millennium Development Goals.

**SDGs** Sustainable Development Goals.

**Player** Any individual, organisation or group involved in a geographical issue, as either the cause, suffering the consequences, or seeking to manage/govern it.

- farmers — can organise themselves into collectives and have greater strength to negotiate as groups, e.g. the Fairtrade movement
- NGOs (e.g. Greenpeace, Oxfam) — can lobby, raise public awareness and fund projects

**Exam tip**

It is important that you develop a view on these issues and consider how others may see them differently from you.

### Do you know?

1 Summarise the main features of Article 1 of the UN Charter.
2 Through which agencies does the UN provide humanitarian assistance around the world?
3 Outline the key elements of the Fairtrade movement.
4 What is meant by the term 'ethical trade'?

# 4.5 The 'global commons' and Antarctica

### You need to know
- the concept of the 'global commons', with specific reference to Antarctica
- the physical context of Antarctica
- the threats Antarctica faces
- how Antarctica is governed, including the role of NGOs

## The 'global commons'

Key points:
- only one-third of Earth's surface is divided into sovereign states — bounded territories where individual authorities have absolute power
- the rest of the planet, and beyond, is viewed as a 'global commons': a space that is open to all of humanity

The designation of a 'global commons' can mean it is:
- governed by global treaty, which, in theory, prevents individual states harming it
- free for all to use and for the common good
- subject to debate as to what each of 'treaty', 'free' and 'common good' actually means

**Key term**

Global commons The Earth's shared resources, such as the deep oceans, the atmosphere, outer space and Antarctica.

# Antarctica

## Geography

Physical aspects:

■ Antarctica is the most southern continent and contains the geographic South Pole

■ it is almost entirely south of the Antarctic Circle and is surrounded by the Southern Ocean

■ its size is estimated to be 14 million km², making it the fifth-largest continent

■ 98% of the land area is covered by ice, which averages almost 2 km in thickness

■ only the most northern reaches of the Antarctic Peninsula are uncovered

■ it has the highest average elevation of all the continents

> ### Exam tip
>
> Note: The AQA specification includes the Southern Ocean as far north as the Antarctic Convergence.

**Figure 33 Physical map of Antarctica**

Climate:

- Antarctica is the coldest, driest and windiest of all the Earth's continents
- around the coasts temperatures are close to freezing in the summer months, and slightly positive in the northern part of the Antarctic Peninsula
- during winter, monthly mean temperatures at coastal stations are between −10°C and −30°C, though temperatures may rise towards 0°C when winter storms bring warm air
- temperatures on the high interior plateau are colder as a result of its elevation, latitude and greater distance from the ocean
- here, summer temperatures struggle to get above −20°C and monthly means fall below −60°C in winter
- the distribution of precipitation over Antarctica is varied, with heavy snow falling near the coast (200 mm) but the interior getting only small amounts
- gales can occur more than 40 days a year — mean speeds can exceed 30 m s⁻¹, with gusts of over 40 m s⁻¹
- strong **katabatic winds**, caused by the flow of cold air off the central plateau, make some coastal sites around Antarctica the windiest places in the world

## Threats

According to the Intergovernmental Panel on Climate Change (IPCC):

- Antarctica is facing the loss of ice from the ice sheet, especially from the Antarctic Peninsula (AP)
- the majority of this ice loss will take place on the **WAIS**
- the WAIS is drained by several ice streams — the movement of which is increasing
- the AP is one of the most rapidly warming places in the world — air temperatures here have increased by 3°C over the last 50 years
- it is also seeing losses from the ice shelves around its shores — a recent large crack in sea ice has appeared in the Ross Ice Shelf

Fishing and whaling:

- fishing, sealing and whaling have long featured in the seas around Antarctica
- fishing became the main economic use of the seas during the 1960s/70s, when rock cod and krill stocks were fished by Russian and Japanese fleets using large trawlers
- over-exploitation has led to depletion — the 'tragedy of the commons'

### Key terms

**Katabatic winds** Winds that blow down valley sides and valley floors.

**WAIS** West Antarctic Ice Sheet.

### Exam tip

Keep up to date with what the IPCC states about climate change in Antarctica.

■ as commercial whaling has declined elsewhere, distant-water whaling vessels have become more prevalent in the Southern Ocean

Mineral resources:

■ mineral mining is prohibited by the Madrid Protocol, an extension of the ATS
■ some countries, including the USA, UK and Japan, have argued against this
■ the ban can be revisited in 2048
■ there have been recent signs that some countries (China, Russia) want to revisit the ban sooner as they have an interest in the mineral wealth

Tourism and scientific research:

■ small-scale tourism began in Antarctica in the 1950s, but has since increased — over 52,000 tourists visited in 2013/14
■ tourism is still relatively small-scale, so could be argued to be sustainable
■ most tourism takes place in summer — this is the breeding season for most of the wildlife, however, so tourism could cause disturbance
■ there is also pressure on the whaling station landing sites, e.g. McMurdo Sound, where the original huts from Scott's expedition in 1912 are located
■ IAATO ensures that tourism is conducted in an environmentally friendly way
■ The BAS also welcomes a small number of visits to some of its stations during the austral summer

## Management

Several treaties and organisations seek to govern Antarctica and its environs.

The Antarctic Treaty — 1959:

■ the ATS banned all forms of military activity, and made it a zone free of nuclear tests and disposal of radioactive waste
■ promoted international scientific research, guaranteeing the rights of all states to establish research stations
■ set aside any disputes over land on the continent

The Madrid Protocol — 1998:

■ an extension of the ATS called the Protocol on Environmental Protection
■ banned mining activities (except for scientific research)

> **Key terms**
>
> **ATS** Antarctic Treaty System, a form of governance, sanctioned by the UN.
>
> **IAATO** The International Association of Antarctica Tour Operators.
>
> **BAS** British Antarctic Survey — a scientific research organisation with seven stations in the area.

The International Whaling Commission:

- seeks to conserve whale stocks by protection, catch limits and size restrictions
- designates whale sanctuaries in the Southern Ocean

The International Whaling Moratorium:

- declared a pause in commercial whaling
- is still in place, although Japan evades it by 'special permit', and Norway and Iceland object to it
- allows 'aboriginal subsistence' whaling in Greenland and Alaska

United Nations Environment Programme:

- the UNEP prepares a regular report on Antarctica for the UN Secretary-General (UNSG)
- a report is presented by the UNSG on the 'Question of Antarctica' every 3 years

NGOs:

- have an active interest in the protection of the Antarctic and the Southern Ocean
- can undertake very little in terms of direct impact
- major involvement has been to monitor threats and ensure that the various protocols are enforced

> ### Exam tip
>
> Research the work of the Scientific Committee on Antarctic Research (SCAR), the Antarctic and Southern Ocean Coalition (ASOC) and the Antarctic Oceans Alliance.

> ### Exam tip
>
> It is important that you develop a view on the issues facing Antarctica and the governance that is in place, and consider how others may see them differently from you.

## Do you know?

1 What is the Antarctic Convergence?
2 How many people live on Antarctica?
3 Outline the flora and fauna found on Antarctica.
4 What are krill and why are they important for the food chain?
5 What is the role of the International Whaling Commission?

# End of section 4 questions

**1** Explain how one transnational corporation (TNC) has contributed to the globalisation of the world's economy.

**2** Outline the concept of the 'shrinking world'.

**3** Discuss the various dimensions of globalisation.

**4** Assess how the practice of outsourcing labour from richer to poorer countries might present both challenges and opportunities for the host countries.

**5** Explain how security is a contributory factor to globalisation.

**6** To what extent is China driving the global economic system to its own advantage and influencing geopolitical events?

**7** Evaluate the impact of TNCs on the global economy.

**8** Explain why there is a need to protect and conserve the 'global commons'.

**9** 'The global economy has moved on from the developed nations of the world; the future of the global economy now lies elsewhere.' To what extent do you agree with this statement?

**10** 'The impact of changing carbon budgets is a much greater threat to Antarctica than the impact of tourism or fishing and whaling.' To what extent do you agree with this statement?

# 5 Changing places

## 5.1 The nature and importance of places

### You need to know

- the concept of place
- insider and outsider perspectives on place
- different categories of place
- the factors contributing to the character of places

## The concept of place

A place:

- can have an objective meaning, such as map coordinates or its **location** on a **GPS**
- is given subjective 'meaning' by people — a **sense of place**
- creates an important basis of life — a 'lived experience'

The concept of place can be summarised as:

place = location + meaning

## Insider/outsider perspectives

### Key terms

**Location** A point in space with specific links to other points in space.

**GPS** Global positioning system.

**A sense of place** The personal feelings associated with living in a place.

### Exam tip

Create in your notes a list of 'places' that are important to you at a variety of scales. They will be useful to support your arguments.

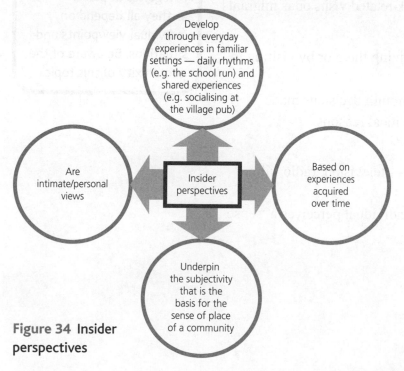

**Figure 34 Insider perspectives**

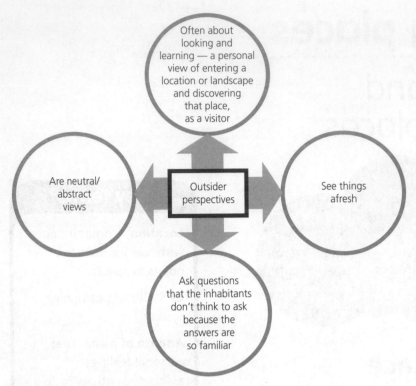

**Figure 35 Outsider perspectives**

# Categories of place

Near places:
- the centre of most lived experiences — most used by 'locals'
- where everyday experiences take place

Far places:
- beyond immediate lived experiences — seen through media, magazines, books, computer contact (business and leisure)
- where people visit as tourists, on work-related visits or as migrants

Experienced places:
- influence people's lives directly by living there or by visiting in person
- influence people's lives indirectly through decisions made there — for example, for work or political reasons

Media places:
- brought to the home through media — television, radio, film, internet, books
- seen through personal 'eyes' — an individual perceives a 'sense of place'

**Exam tip**

Any location can satisfy one or more of these categories of place, as they all depend on individual viewpoints and situations. Be aware of the complexity of this topic.

# Factors contributing to place

Endogenous factors:

■ natural characteristics — geology, altitude, slope angle, coastal or inland

■ demographic — number of inhabitants, their ages, genders and ethnicities

■ socio-economic characteristics — types of household (young adults, families with children, retired people), income levels, levels of education, types of employment (e.g. skilled, manual, professional)

■ cultural factors — religious groups, local traditions

■ political factors — type of local, regional or national government; local groups such as residents' associations and campaigning groups

■ the built environment — the ages/styles of buildings and their building materials; flats, terraced, semi-detached or detached housing; building density

Exogenous factors:

■ features associated with globalisation — clothing shops, fast food outlets, places of employment, social media

■ evidence of multi-ethnicity caused by inward migration

> ## Key terms
>
> **Endogenous factors** The factors that originate from within, i.e. internally.
>
> **Exogenous factors** The factors that originate from without, i.e. externally.

# Your studies of 'place'

At the heart of this unit of work are the two detailed case studies (see 5.4 Place studies, page 93):

■ one local place where you live or study

■ one further, contrasting place

When using the following two topics, 5.2 Relationships and connections and 5.3 Meaning and representation:

■ ensure that you appreciate how the factors they mention affect continuity and change in your chosen places

■ reflect on how your life and the lives of others have been, and are, affected by such continuity and change in your chosen places

> ## Do you know?
>
> 1 Distinguish between 'space' and 'place'.
>
> 2 The four categories of place owe their origin to the ideas of Yi-Fu Tuan. He used the phrase 'field of care'. What is a 'field of care'?
>
> 3 With reference to one example in the media, examine its relationship with 'place'.

# 5.2 Relationships and connections

## Shifting flows

### People

Shifting flows of people help shape a place profile, due to:

■ migration — people move into a place or they move away; these flows of people can change the demography of a place

■ life-cycle stage:

☐ an area can receive an influx of young adults, e.g. students or young professionals living away from home

☐ an area can be affected by an influx of commuters due to accessible railway lines and stations

☐ an area can receive an influx of retired people, e.g. some UK coastal towns

### Resources

Resources can shape a place profile:

■ the availability of a mineral resource can lead to the establishment of a mining community

■ if that resource runs out or is no longer required, the place undergoes change, although evidence of its previous profile persists

■ the rise in electronic communication has given great significance to the connectivity a place has via the internet and mobile technology, e.g. remote rural locations

### Money and investment

Money and investment help shape a place profile, with:

■ funds provided by governments, TNCs and local authorities (LAs)

---

## Key terms

**Migration** A permanent change in where a person lives — can be internal or international.

**Life-cycle stage** Describes the age and family status of a person, such as young adult, married with children or retired.

**Resource** Requires a broad definition — anything that people can make use of.

## Exam tip

The specification states that you should focus *either* on changing demographic and cultural characteristics *or* on economic change and social inequalities. These overlap, however. It is advised that you do *not* categorise in this way. Study all elements and use the better set of information for each question as it arises.

- spending on characteristics of places such as transport infrastructure, education, health and environment
- small-scale funds from individuals or groups of individuals can result in the creation or modification of businesses — e.g. attempts to resist **clone high streets**

# External forces

A number of large-scale **agents of change** can affect a place profile — governments, TNCs and international/global institutions. They often act together to change the character of a place, and so it is difficult to separate out their relative impacts.

- governments — invest in the infrastructure of a place, thereby enabling change; LAs give enticements to others to invest in the locality
- TNCs — provide investment that gives them the greatest return, e.g. a factory or a hotel complex
- the EU — has provided vast sums of money to regenerate declining areas across the UK

# Past and present connections

Key points:

- places do not exist in total isolation from the rest of the world — over the time a place has existed it has been connected to other places (regionally, nationally and internationally)
- present-day connections — e.g. journey-to-work patterns, holiday destinations, sources of food and clothing — are extensive
- past connections can be evidenced by the ethnic composition of the population, old buildings, disused docks, public spaces, street names, and even the place's name

> ## Key terms
>
> **Clone high street** One dominated by chain shops that can be found all over a country; few local, independent businesses survive.
>
> **Agents of change** Individuals, groups, corporations, institutions, media and governments that drive change, either intentionally or unintentionally.

> ## Exam tip
>
> Consider the place where you live. Think of public spaces, large buildings and street names that connect with the past. You can use these to illustrate exam answers.

> ## Do you know?
>
> 1 Give examples of how resources influence a place profile.
> 2 Explain the link between globalisation and clone high streets.
> 3 Explain how different levels of government can impact on a place.
> 4 The word 'palimpsest' is used when considering the past connections of a place. What does it mean?

# 5.3 Meaning and representation

## You need to know

- how people perceive and form attachments to places, including the way everyday place-meanings are connected with different identities, perspectives and experiences
- how external agencies attempt to influence or create place-meanings
- how places may be represented in a variety of different artistic forms, which often give contrasting images to those presented by more formal methods
- how past and present processes of development can be seen to influence the socio-economic characteristics of places

## Attachments to place

### Perception

Key features:
- people perceive places in different ways — our **identity** influences our perceptions
- the way we see a place may not be the same as the way other people see it
- age, gender, ethnicity and sexuality can impact on perceptions of a place

Age:
- perceptions of a place often change throughout our life-cycle
- young adults may prefer to live in locations where work, shops and leisure facilities are close by
- people with young families may desire more space (a garden, a park), with access to a nursery or school
- older people may prefer a more secluded place

Gender:
- perhaps not as influential in developed societies as in the past, though stereotypical attitudes may prevail regarding work vs home
- some workplaces are still dominated by one gender — e.g. mining, nursing

### Key term

**Identity** An assemblage of personal characteristics such as gender, sexuality, race and religion.

### Exam tip

Consider a town/city you know well. Think about the location of places where different groups of people might live or meet, and their reasons for doing so.

■ the perception of 'safety' (or lack of it) in some urban areas still impacts on gender — e.g. train carriages at night

Ethnicity and sexuality:

■ places with concentrations of particular ethnicities have developed for many years

■ the character of the place may have changed to reflect the place of origin of the ethnic group — e.g. 'Chinatowns' in cities

■ shops and services (e.g. religious) are also influenced by these groups

■ some places in cities have concentrations of LGBTQ+ groups

■ for all of the above, the sense of belonging is enhanced and provides security

## Attachment

Key features:

■ attachment can come with long-term association with a place — based on memories

■ people can still feel attached to a place even though they no longer live there

■ for some, attachment to a place is more subjective — a 'homeland' with a strong sense of identity, but no geographical or political basis

# Influencing and creating place-meanings

Key points:

■ the impact of various agents of change has been mentioned earlier (see page 89)

■ these organisations attempt to manage, or manipulate, perceptions to create a place-meaning

See Figure 36.

> **Key term**
>
> Belonging  A sense of being part of a collective identity.

> **Exam tip**
>
> Perception is an important influence on the meanings people give to places. Make sure you have real-world examples to support your knowledge and understanding.

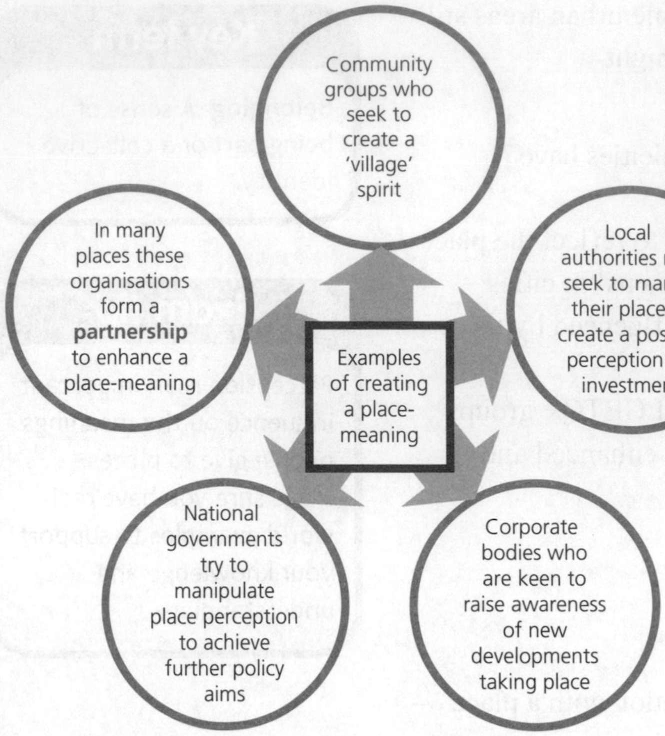


**Figure 36** Examples of creating a place-meaning

## Key terms

**Partnership** Non-governmental and governmental organisations working together to develop or regenerate an area and give it a new identity.

**Representation of place** The cultural practices by which people interpret and portray the world around them and present themselves to others.

**Geospatial data** Locational information about a place — geographic information systems (GIS) are often used.

# Representation of place

Artistic methods:

- several agencies make use of our imagination to influence how we see a place
- advertising agencies combine visual and written imagery to enhance the settings of a place
- tourist boards may select aspects of a place that fit a desired perception of that place
- some **representation of place** may be informal but no less powerful — novels, poems, songs, the visual arts and other diverse media (television, film, video, photography) all 'bring alive' different places
- in these cases we may struggle to match the fictional with the factual

Formal methods:

- more data about places are now collected, stored and analysed than ever before
- in many countries the most effective formal representation of a place is its census
- there has been a dramatic increase in the quantity and quality of **geospatial data**
- many government agencies maintain websites that present formal representations of places

## Exam tip

Consider how twenty-first-century advertisers entice you to tourist locations.

■ formal representations offer rational perspectives of a place profile, such as numbers of people living in a place, their ages, genders and educational qualifications

■ they are limited, however, in their ability to indicate 'lived experience' aspects of a place profile

# Past and present processes

Key points:

■ there is an uneven spread of resources, wealth and opportunities within and between places

■ **social inequality** tends to focus on differences in access to housing, healthcare, education and employment

■ in the UK these are combined to give an **IMD** (see page 96)

■ several factors determine social inequality; some have operated in the past, and some operate in the present:

  ☐ levels of income: in general terms, the higher the level of income, the higher the standard of living; unemployment increases social inequality, and both are linked to past industrialisation or deindustrialisation

  ☐ housing: affordability of housing is crucial, either for rent or purchase; in some rural areas demand from second-home ownership has caused problems for young adult residents

  ☐ health: ill-health is associated with sub-standard housing, poor diet and unhealthy lifestyles; access to healthcare is crucial

  ☐ education: access to education, both primary and secondary, is an issue for the developing world; in the UK there are perceived differences between public and private, inner-city and suburban

## Key terms

**Social inequality** The uneven distribution of opportunities and rewards for different social groups, defined by factors such as age, gender, class, sexuality, religion or ethnicity.

**IMD** Index of Multiple Deprivation.

## Exam tip

Consider the advantages and disadvantages of methods of measuring social inequality, such as income, education and health.

## Do you know?

1 Give an example of an attachment to a place but with no political basis.

2 Why do informal representations of place need to be interpreted with care?

3 Why is housing a good indicator of social inequality?

# 5.4 Place studies

This part of the specification requires you to apply the knowledge acquired through engagement with the previous content. Use this knowledge to enhance your understanding of the way your life and those of others are affected by continuity and change in the nature of places. Sources must include **qualitative** and **quantitative data** to represent places in the past and present.

You must undertake two place studies:

1 **Local place study:** exploring the developing character of a place local to your home or study centre.
2 **Contrasting place study:** exploring the developing character of a contrasting and distant place.

Both place studies must focus on people's lived experience of the place in the past and at present, and *either* changing demographic and cultural characteristics *or* economic change and social inequalities.

## Advice

### What is a 'local place'?

It is a place:
■ that is familiar to you
■ with which you have some personal experience
■ that is within a few miles of your school

### What is a 'contrasting place'?

It is a place:
■ that is 'distant' from the local place in terms of character
■ with which you are unfamiliar
■ that may also be geographically distant, e.g. located in a different country to the local place study or in an unfamiliar part of the UK (or your home country)

It is equally acceptable for the contrasting place to be located near to the local place as long as it provides the necessary contrast.

The contrasting place must show significant contrast to the local place in terms of each of:
■ economic development
■ population density
■ cultural background
■ systems of political and economic organisation

## Key terms

**Qualitative data** Non-numerical data such as photographs and sketches; may involve the collection of opinions, perspectives and feelings from questionnaires and interviews.

**Quantitative data** Data in numerical form, which can often be placed into categories and analysed statistically.

## Exam tip

The specification states that you should focus *either* on changing demographic and cultural characteristics *or* on economic change and social inequalities. These overlap, however. It is advised that you do *not* categorise in this way. Study all elements and use the better set of information for each question as it arises.

## Exam tip

The idea of you having some experience of the area is important. For this reason, a field study centre location would not be suitable for the local place study (unless it is located in close proximity to your school).

# 'Local place' tips

Choose:

- a place with access to a good supply of data
- an urban place that has undergone or is undergoing change — this could be socio-economic (economic decline or regeneration, new housing estate, near an industrial estate or out-of-town shopping centre) or demographic/cultural (in- or out-migration)
- a rural place that has experienced counterurbanisation, depopulation or a contested landscape (new housing estate, second homes, infrastructure projects such as HS2)

> ### Exam tip
> How large should a 'place' be? One you could walk around on a one-day fieldtrip; one with fewer than 20,000 people; a ward of a town.

# 'Distant place' tips

Choose a place:

- that has good availability of data and the different sources of information
- where charities and non-governmental organisations are useful sources of information
- with which there is a possible twinning arrangement with a school or where you can make use of video-calling to talk to people
- you have visited, or will visit

# Techniques

## Maps

Key features:

- start with a 1:25,000 Ordnance Survey map
- look at the physical geography of the area: relief, height of the land, aspect, drainage
- how have humans impacted the area?
- what are the main land uses?
- you could see how the area has changed by looking at older OS maps
- look for changes in the size of the place, types of land use and infrastructure
- you could carry out fieldwork to map land use within the area

## Literary sources

Use books, atlases and newspapers alongside maps to provide historical information about places. Produce a potted history of the place.

## Geospatial data

Consider using online sources for socio-economic data:

■ Office of National Statistics (ONS):
  **www.neighbourhood.statistics.gov.uk**
■ DataShine Maps (based on the 2011 census):
  **http://datashine.org.uk**
■ Consumer Data Research Centre (CDRC) maps:
  **http://data.cdrc.ac.uk**
■ IMD: **http://apps.opendatacommunities.org/showcase/deprivation**

**Exam tip**

Note: The more detailed DataShine and CDRC maps present data by Super Output Area (SOA), whereas the ONS mapping tool uses data aggregated by ward.

## The Index of Multiple Deprivation

Key features:

■ the IMD is published by the Department for Communities and Local Government and informs national and local government decision making
■ ranks the SOAs across England according to a combination of seven domains of deprivation: income, employment, education, health, crime, barriers to housing and services, and living environment
■ each of these domains is based on a further number of indicators — 37 in total
■ each indicator is based on the most recent data available, although in practice most indicators in the 2016 data, for example, relate to the tax year ending April 2014
■ uses smaller parts of the SOAs, called Lower SOAs (LSOAs) — each contains about 1500 residents or about 650 households. This allows the identification of small pockets of deprivation

**Key term**

SOA Super Output Area — a basic area of the UK census.

The deciles shown on the IMD maps are produced by ranking 32,844 LSOAs and dividing them into ten equal-sized groups. Decile 1 represents the most deprived 10% of areas nationally, and decile 10 the least deprived 10%. When interpreting these data, note:

■ the rank of the deciles is relative: they simply show that one area is more deprived than another but not by how much
■ you will see large areas of colour, often the same or similar shades. Note that these show areas and not numbers of people living there
■ the data shown by such neighbourhood-level maps provide a description of an area as a whole and not of individuals within that area
■ deprivation, not affluence, is mapped

## Other qualitative sources

You must use one or more of the following for your two places:

- photographs — could be from online sources such as Flickr or Google Earth
- postcards
- paintings
- written documents — novels, old newspaper accounts, poems
- marketing literature
- films and/or videos
- songs
- interviews or questionnaires
- social media sources — e.g. Twitter, Instagram, TripAdvisor, blogs

### Exam tip

Note that 'lived experience' is an essential part of your place studies. Interviews and social media sources are ideal for this.

### End of section 5 questions

1 Choose a place you know well that has been represented in some form of media. How well did the media illustrate the 'true' characteristics of the place?

2 Explain how shifting flows of investment can change either the demographic characteristics or the socio-economic characteristics of a place.

3 Distinguish between outsider and insider perspectives of a place.

4 For one place you have studied, identify a qualitative source (for example a painting, an old photograph, a newspaper article, an interview or a song) and discuss how this source helped you understand and perceive that place.

5 In your study of a place, you will have come across different perceptions of that place held by different people. Describe two of these perceptions and discuss the extent to which they match the quantitative data on that place.

6 Distinguish between endogenous and exogenous factors in the study of the character of places.

7 How can a place be seen to develop in layers of time?

8 Assess the extent to which the experiences of people living in a place that you have studied have been affected by the development of the area's infrastructure.

9 Compare and contrast the usefulness of different visual sources in helping you understand the local place you have studied.

10 Evaluate how far the lived experiences of different groups in a place that you have studied have led to different place identities.

# 6 Contemporary urban environments

## 6.1 Urbanisation

### You need to know

- the importance of urbanisation in human affairs
- the global pattern of urbanisation
- the variety of processes that have operated, and continue to operate, within urban areas
- how megacities and world cities have emerged
- urban policy and regeneration in Britain since 1979

**Urbanisation** occurs when:

- rural to urban migration is greater than urban to rural migration
- life expectancy and natural population increase are greater in an urban area

Resulting issues that are important:

- the development of **primate cities**, where there are marked differences between the economic core and peripheral regions
- cities in the developing world may not be able to provide enough housing for all of the migrants so they are faced with the alternatives of sleeping on streets or building **shanty towns**
- these in turn create pressure on refuse collection, health provision, education, police and fire services, power supplies, transport systems and sewage disposal
- the number of jobs is inadequate, leading to unemployment, under-employment and a large **informal sector**
- many cities in the developed world have faced periods of economic decline, leading to a range of strategies aimed at improving them

## Global pattern

Key features:

- between 1945 and 2015, the proportion of the global population that is urban increased from one-third to one-half — by 2050 it is expected to be two-thirds

### Key terms

**Urbanisation** The process whereby the proportion of people living in towns and cities increases.

**Primate city** Where one city in a country dominates the city size distribution — it is more than twice the size of the second-largest.

**Shanty town** An illegal settlement within a city that contains cheap, often hand-built, houses.

**Informal sector** Unregulated and unstructured employment.

- the number of people who live in urban areas increased from 0.7 billion to 4 billion in this time period, with a projection of 6.3 billion for 2050
- North America and Latin America are the most urbanised regions, with 80% or more of their populations living in urban settlements
- Europe (73% urban in 2015) is expected to be more than 80% urban by 2050
- Africa and Asia remain mostly rural, with 40% and 48% of their respective populations living in urban areas — both regions are projected to urbanise faster than other regions, however, reaching 56% and 64% urban respectively by 2050

# Urban processes

## Deindustrialisation

Occurs:

- when there is an absolute decline in the importance of manufacturing in the economy of a country — there is a fall in its contribution to GDP
- where there has been an over-reliance on traditional heavy industries, such as iron and steel, chemicals, shipbuilding and textiles
- due to strong overseas competition from areas where new technology and less unionised practices have been adopted

> **Exam tip**
>
> Note that a wide range of factors (economic, social, technological, political and demographic) influences each of these processes. Each factor plays a role, but in varying degrees.

## Decentralisation

Occurs:

- when there is an outward movement of population and industry from established central urban areas towards the suburbs or to smaller urban centres
- where there is encouragement by government agencies trying to spread investment and development from the central area of a city towards the periphery
- due to the negative aspects of higher crime, noise and pollution levels, and higher land costs found in central locations

## The rise of the service economy

Key points:

- deindustrialisation and decentralisation have both encouraged the rise of the service economy in developed world cities

- there has been an increase in employment in education, health, public transport, retailing, local government, banking and finance
- this sector employs the most people in a developed country

**Exam tip**

Note that these three processes also operate at a national scale. The following processes are more at a city scale.

## Suburbanisation

See Table 17.

**Table 17 Characteristics, causes and effects of suburbanisation**

| Characteristics | Causes | Effects |
|---|---|---|
| Extensive areas of housing built on the edges of major cities<br><br>More recently there has been the development of new housing areas on brownfield sites, leading to the infilling of some large private gardens and other open spaces, such as school playing fields | The growth of public transport systems and the increased use of the private car<br><br>The presence of railway lines and arterial roads has enabled relatively wealthy commuters to live some distance away from their places of work but in the same urban area | Outward growth of urban development that has engulfed nearby villages and rural areas — urban sprawl<br><br>The urban edge, where more land is available for car parking and expansion, becomes the favoured location for new offices, factories and shopping outlets<br><br>Designation of green belts |

## Counterurbanisation

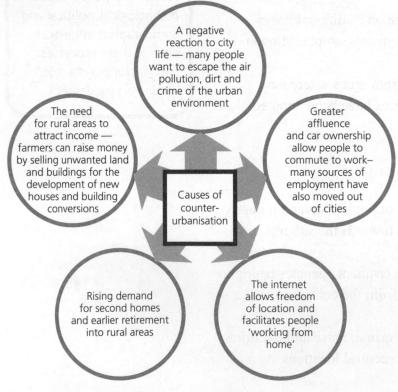

**Figure 37 Causes of counterurbanisation**

Effects include:
- an increase in the use of commuter railway stations, with large areas for car parking
- above-average increases in the value of houses

**Key terms**

**Suburbanisation** The movement of people from living in inner-city areas to living on the outer edge of a city.

**Brownfield site** Previously developed land that is being redeveloped — including urban green space.

**Green belt** Land around a city that is left rural and undeveloped — an attempt to prevent urban sprawl.

**Counterurbanisation** The movement of people from larger urban areas into smaller urban areas or rural areas — leapfrogging the rural–urban fringe.

- the growth of new small housing estates in old villages
- problems for young adult locals when they want to find homes to rent or buy

# Urban resurgence

Factors causing **urban resurgence**:

- regeneration — the investment of capital and ideas into a rundown city area to revitalise and renew its economic, social and/or environmental condition
- rebranding — the process of regenerating a city's economy and physical fabric, as well as projecting a new, positive urban image to the wider world
- gentrification — housing improvement by people rather than by organisations; associated with a change in neighbourhood composition in which low-income groups are displaced by more affluent people, usually in professional or managerial occupations (see page 108)

General effects include:

- the reuse of old buildings for a new purpose — housing, offices, hotels
- positive multiplier effect, with more investment being attracted
- vibrant city-centre activities — the '24 hour' city
- displacement of lower-income people, often with increasing levels of inequality

# Megacities and world cities

Megacities:

- over 30 existed in 2015, with Tokyo, the largest, having over 35 million inhabitants
- the number of **megacities** is expected to increase to 40 by 2030
- recent megacity growth since 2000 has been centred on Asia, especially India and China
- Africa has relatively few megacities, although many African cities are growing rapidly

World cities:

- unlike megacities, **world cities** are mainly located in the developed world — the four top-ranking world cities are New York, London, Paris and Tokyo
- are leading business centres and the preferred headquarter locations of leading TNCs — they are also global service centres, specialising in advanced producer services such as

> ## Key terms
>
> **Urban resurgence** The movement of people back into urban areas, particularly the inner-city — often associated with urban regeneration, rebranding schemes or gentrification. It is also associated with the move towards sustainable urban communities.
>
> **Megacity** A city with more than 10 million inhabitants.
>
> **World city** A city that has 'global' influence as a major centre for finance, trade, politics and culture.

> ## Exam tip
>
> Know some specific examples of each of these urban-resurgence processes.

> ## Exam tip
>
> Know the names and locations of some of the largest cities in the world — use an atlas.

finance, banking, accounting, management consultancy, law and advertising
- are major telecom, information and transport hubs
- are magnets for highly educated, skilled workers, and are home to world-class universities
- have a political and cultural dimension, housing foreign embassies, consulates and international organisations, hosting international sporting events, and supporting a wide range of performing arts venues as well as museums, galleries and restaurants

# British urban regeneration policy

See Table 18.

Table 18 Major forms of urban policy and regeneration in Britain (1979–present)

| Policy | Date | Main features |
| --- | --- | --- |
| Pump-priming | 1979–98 | Autonomous Urban Development Corporations (UDCs) and Enterprise Zones were given responsibility for infrastructure renewal and derelict site remediation in order to attract private sector investment. UDCs were free from many planning regulations; local and national government involvement was minimal. |
| Public–Private Partnerships (PPPs) | 2000–10 | Regional Development Agencies (RDAs) worked to combine private and public investment to regenerate key sites in cities. Flagship and landmark buildings were used to 'kick start' investment. |
| Local Enterprise Partnerships (LEPs) | 2010–present | The coalition government abolished most RDAs in 2010. Since then regeneration has been largely led by private investment, often by retail (e.g. Westfield schemes) and private housing developments. The scale of regeneration is much more local. Local councils act merely as advisors and facilitators. |

## Do you know?

1 The continental pattern of urbanisation was provided earlier. Examine the variations that exist *within* some continents.
2 Give one economic, one social and one environmental impact of deindustrialisation.
3 Name some well-known and some less well-known megacities.
4 How do you measure a world city's brand?
5 Identify ways in which the UK national government has assisted urban regeneration.

# 6.2 Urban forms and associated issues

**You need to know**

- the general characteristics of urban forms in contrasting settings
- the spatial patterns of land use, inequality, segregation and cultural diversity in contrasting urban settings, and the issues associated with them
- the strategies to manage the above issues
- the characteristics of a range of new urban landscapes

## General characteristics of urban form

Key features:

- over the last hundred years sociologists, economists and geographers have developed a number of theories (models) seeking to explain **urban form**
- they considered the physical and human factors that have shaped urban areas
- they attempted to describe the spatial pattern of land use, economic inequality, social segregation and cultural diversity, as well as to consider the factors that have influenced the development of these patterns
- megacities have tended to grow rapidly in the developing world, with haphazard growth creating challenges for authorities in providing services for inhabitants
- world cities, being largely in the developed world, have tended to develop in a more organised and planned manner

## Urban form in the developed world

Key points:

- physical factors were important in historical times — towns were built alongside rivers, on flat land, and often around a river crossing (bridging point)
- as time progressed, economic factors involving land values became more important — the **PLVI** being crucial (Figure 38)
- the PLVI became the economic core of the urban area — the **CBD** — with other land uses (industrial and residential)

### Key terms

**Urban form** The arrangement of land use in urban areas — shape and spatial pattern.

**PLVI** Peak land value intersection.

**CBD** Central business district.

### Exam tip

Three of the best-known models of urban form for the developed world are: the concentric zone model (Burgess), the sector model (Hoyt) and the multiple nuclei model (Harris and Ullman). Research one of these models.

occurring as determined by the price of land as distance increases from the CBD (bid rent theory — Figure 39)

■ in recent decades suburban office and shopping areas have resulted in secondary land-value peaks

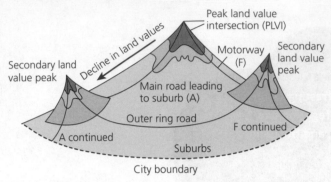

**Figure 38** Land values for a typical city in a developed country

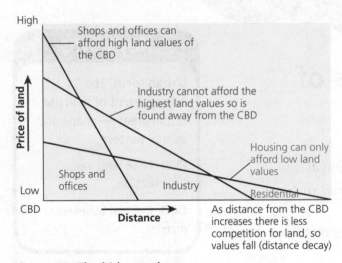

**Figure 39** The bid rent theory

# Urban form in the developing world

Ford presented a model of a developing world city based on Latin American cities (Figure 40):

■ cities are built up around a central core consisting of a CBD and market area

■ from this extends a commercial spine flanked by an elite residential sector

■ a separate industrial sector/park develops on the edge of the city

■ middle-class housing locates close to the elite housing sector and the **periférico**

■ some gentrification to protect historical quarters and some consolidation of housing (zone of maturity) take place near the centre

■ concentric zones of housing decrease in quality with distance from the centre

■ squatter settlements (**favelas**) progressively develop in the **periferia**

■ over time these neighbourhoods become more stable and obtain more services

### Key terms

**Periférico** An outer ring road around a Latin American city.

**Favela** A shanty town in Latin American cities.

**Periferia** The outer parts of a Latin American city.

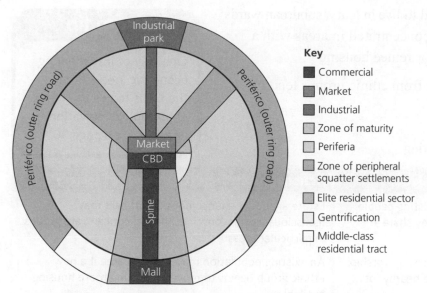

**Figure 40** Ford's model of Latin American cities (1966)

Key
- ■ Commercial
- ■ Market
- ■ Industrial
- ☐ Zone of maturity
- ☐ Periferia
- ☐ Zone of peripheral squatter settlements
- ☐ Elite residential sector
- ☐ Gentrification
- ☐ Middle-class residential tract

## Exam tip

Examine these issues in the context of the two case studies of contrasting urban areas you have to undertake (see 5.4 Place studies). Make sure you look at patterns of economic and social wellbeing and at the impact of environmental conditions for these two cities.

# Social and economic issues

## Issues

In general terms the issues are:

- **economic inequality**:
  - ☐ exists in terms of access to job opportunities, education, housing and basic services such as water and sanitation
  - ☐ knock-on impacts are poorer health, higher unemployment and a lack of social mobility for those at the poorest end of the spectrum — they get stuck in a cycle of poverty from which it is hard to escape
- ethnic communities become isolated from wider society as they maintain their **cultural diversity** (origin, language, beliefs) and have limited interaction with others
- consequently, both economic inequality and cultural diversity can lead to **social segregation**

## Segregation in the UK

Examples of social segregation in the UK:

- **gated communities** can be found adjacent to '**sink estates**'
- in rural areas, successful, prosperous commuter villages might be only a few miles from less attractive rural villages suffering population decline and service deprivation
- expensive riverside property in London bought by wealthy European and Arab immigrants

## Key terms

**Economic inequality** The difference in levels of income and living standards in an area.

**Cultural diversity** Variations in ethnicity or cultural values within a society.

**Social segregation** When different groups of people live apart from each other based on wealth, ethnicity, religion or age.

**Gated communities** Wealthy residential areas that are fenced off and have security gates and entry systems.

**Sink estates** Social housing estates that are the least desirable to live in. They tend to house the lowest-income, and most in need, residents.

- wealthier white British people tend to live in leafy, suburban wards
- lower-income ethnic groups are concentrated in areas with a large amount of social (council) or rented housing

There are other reasons why people from ethnic groups tend to cluster (Table 19).

**Exam tip**

One way of investigating economic inequality in the UK is to use the Index of Multiple Deprivation, referred to on page 96.

**Table 19 Factors behind ethnic clustering**

| Actions and attitudes of the ethnic group | Actions and attitudes of the rest of society |
|---|---|
| New immigrants tend to live close to existing people from the same ethnic group, because they share a common language and experiences | Estate agents or council housing officers may (consciously or unconsciously) help concentrate groups in particular areas |
| Ethnically specific services — shops, places of worship, faith schools — encourage others to live nearby for convenience | An existing population may leave an area if a new ethnic group begins to move in, making more housing available |
| A view of 'safety in numbers' and stronger community ties if people live close together | Prejudice (e.g. in jobs) prevents some ethnic groups gaining higher incomes to enable them to move away |

## Segregation in the developing world

Key features:

- the consequences of inequality and cultural diversity are easily identifiable — the city centre core is the centre of wealth, and the periphery is the area of poverty
- there are pockets of wealth dotted in other parts of a city, often 'gated', as in the developed world
- it is estimated that nearly a billion people live in slums or favelas around the world, and this number is increasing — this will deepen the social, economic, political and spatial inequality that is already prevalent

# Strategies

Various governments (national and local), NGOs and individuals (e.g. the Gates Foundation) seek to manage the above issues. In general, this involves:

- managing socio-economic issues, e.g. improved provision of schools, enforcing a living wage, giving access to affordable housing, greater provision of public transport
- dealing with social variations, e.g. increasing the availability of clinics, and health education programmes involving access to sports and leisure facilities
- reducing segregation through legislation on anti-racism and employment rights; combating discrimination and prejudice; encouraging greater political involvement of different cultural groups

**Exam tip**

Examine some of these strategies in the context of the two case studies of contrasting urban areas you have to undertake. Make sure you look at strategies that deal with both economic inequality and cultural diversity.

■ dealing with issues of cultural diversity, e.g. providing English lessons or bilingual literature; hospitals catering for specific illnesses; schools altering their curricula and holiday patterns to cater for different ethnic groups

# New urban landscapes

## Town centre mixed developments

Involves the mixing of retail with:

■ a range of leisure facilities — cinemas, theatres, restaurants and wine bars
■ the provision of public gardens and squares
■ new office and conference facilities
■ the construction of up-market residential areas — often involving gentrification of old buildings

## Cultural and heritage quarters

Often form part of redevelopment schemes, involving one or both of:

■ the presence of a distinctive cultural identity or product — e.g. Chinatown, media industries, the arts, Victorian industrial goods
■ a strong historical connection with an identity or product — e.g. dockland, jewellery

## Fortress developments

A recent trend in urban design is one of 'defensible space', at a variety of scales (Figure 41).

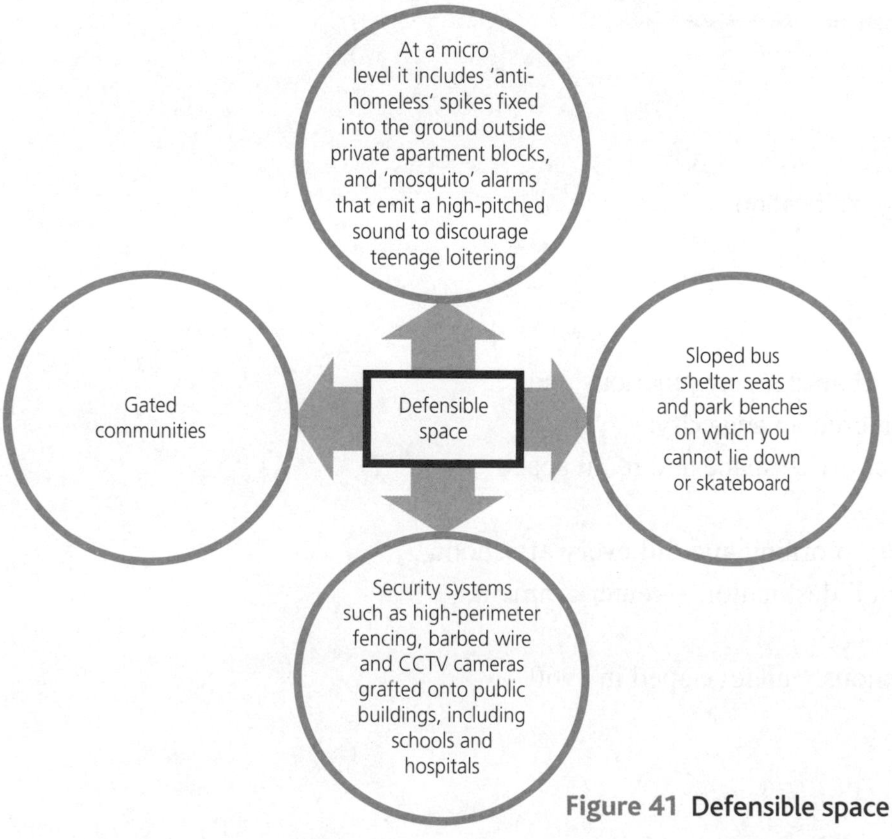

> **Exam tip**
>
> Be able to refer to named examples for each of these new urban landscapes.

At a micro level it includes 'anti-homeless' spikes fixed into the ground outside private apartment blocks, and 'mosquito' alarms that emit a high-pitched sound to discourage teenage loitering

Gated communities

Defensible space

Sloped bus shelter seats and park benches on which you cannot lie down or skateboard

Security systems such as high-perimeter fencing, barbed wire and CCTV cameras grafted onto public buildings, including schools and hospitals

**Figure 41 Defensible space**

# Gentrified areas

Gentrification (Figure 42) involves the:

- rehabilitation of old houses on an individual basis, a process openly encouraged by estate agents, building societies and local authorities
- prosperity of residents becoming greater, leading to an increase in the number of bars, restaurants and personal services
- creation of local employment opportunities in design, furnishings and decoration

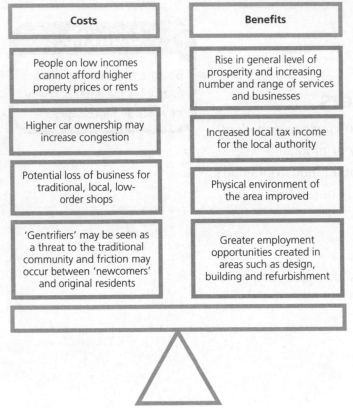

**Costs**

| People on low incomes cannot afford higher property prices or rents |

| Higher car ownership may increase congestion |

| Potential loss of business for traditional, local, low-order shops |

| 'Gentrifiers' may be seen as a threat to the traditional community and friction may occur between 'newcomers' and original residents |

**Benefits**

| Rise in general level of prosperity and increasing number and range of services and businesses |

| Increased local tax income for the local authority |

| Physical environment of the area improved |

| Greater employment opportunities created in areas such as design, building and refurbishment |

**Figure 42 Costs and benefits of gentrification**

# Edge cities

These have developed near suburban freeway junctions and airports. For a place to be considered an **edge city**:

- the area must have substantial office space and retail space (a large shopping mall)
- the population must rise every morning and fall every afternoon
- the place must be a 'single end' destination — entertainment, shopping and recreation
- the area must have been previously undeveloped in 1960

# The post-modern western city

Buildings in **post-modern cities** are:

- more varied in design — e.g. curved, triangular, peaked, with greater amounts of exterior ornamentation
- often centrepieces of artistic and cultural quarters
- often designed around the community who live and work there — giving more power and control to locals rather than being imposed by government

## Do you know?

1 Choose one model of urban form and summarise its main features.

2 Choose one city you have studied and describe the economic inequality that exists there.

3 Choose one urban area you have studied and explain the changes in cultural diversity taking place there.

4 Give some examples of post-modern developments in cities.

# 6.3 Urban climate

## You need to know

- the impact of urban forms on local climate and weather
- specific impacts — on temperature, precipitation, wind and air quality
- strategies to deal with urban air pollution

Key features:

- cities create a 'climatic dome', within which the weather is different from the surrounding rural areas in terms of temperature, relative humidity, precipitation, visibility and wind speed
- for a large city, this dome may extend upwards for up to 250–300 m and its influence may extend tens of kilometres downwind
- within the urban dome, two levels can be recognised — below roof level there is an urban canopy with processes acting in the space between buildings; above this there is a layer with characteristics governed by the nature of the urban surface

# Temperature

The UHIE (Figure 43) is the product of:

- anthropogenic heat sources that include warmth given off by people, machines, heating systems, air-conditioning systems, industrial processes and cars
- multiple reflections of insolation from tall buildings, especially those with high levels of glass
- urban surfaces (concrete, bricks, tarmac) tend to have a lower albedo, which enables them to absorb more insolation; the higher heat capacity of urban surfaces allows them not only to absorb the heat but also to store it
- this heat is then released slowly when the air cools at night
- efficient drainage of the urban surface removes surface water quickly — there is less capacity for evaporation to take place, with its associated cooling effect
- there is less vegetation, which would cool the air by transpiration
- above many cities there is a dome of particulate and $NO_2$ pollution — this allows the short-wave radiation from the Sun into the atmosphere but then absorbs and reflects the outgoing longer-wave radiation, preventing its escape
- often increased cloud over an urban area also reflects outgoing radiation back to the surface

> ## Key terms
>
> **Urban heat island effect (UHIE)** The zone around and above an urban area that has higher temperatures than the surrounding rural areas.
>
> **Insolation** Incoming solar radiation.
>
> **Albedo** The amount of solar radiation (insolation) that is reflected by the Earth's surface and the atmosphere.

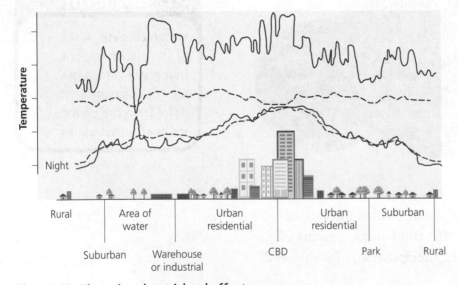

**Key**
— Surface temperature (day)
--- Air temperature (day)
— Surface temperature (night)
--- Air temperature (night)

**Figure 43 The urban heat island effect**

Other points:

- initially, the UHIE was studied because of its warming impact on cities in winter
- scientists are now studying its effect in summer, when potentially lethal heat waves can occur

# Precipitation, thunderstorms and fogs

Precipitation and thunderstorms:

- urban areas have up to 10% more rainfall than surrounding areas
- rain lasts for longer and intense storms with thunder are more common
- the increase in rainfall is because:
  - ☐ the UHIE causes relatively low atmospheric pressure and convectional uplift and consequent convectional rainfall, which also increases the number of thunderstorms and the intensity of the rainfall
  - ☐ the presence of high-rise buildings can cause turbulence, which leads to uplift of air
  - ☐ particulate pollution means that there is an increased number of **hygroscopic** nuclei present in urban air

Humidity and fogs:

- relative humidity is lower in cities than in surrounding rural areas (fewer water bodies, lower rate of evapotranspiration, more rapid runoff of water)
- there is also a greater concentration of airborne particulates that act as condensation nuclei
- this combination means that there is a much higher incidence of fog in urban areas, particularly under **anticyclonic** conditions
- fogs are more frequent in winter and 30% more frequent in summer than in surrounding rural areas
- fogs are most common near urban rivers

# Wind

Key points:

- the surface area of cities is uneven due to the varying heights of the buildings
- buildings tend to exert a powerful frictional drag on air moving over and around them
- this creates turbulence — rapid and abrupt changes in both wind direction and speed
- average wind speeds are lower in cities than in the surrounding rural areas, and they are also lower in city centres than in suburbs
- although high-rise buildings slow down upper air movement, they also channel air into the 'canyons' between them to create the **Venturi effect**

## Key terms

**Hygroscopic** Water-attracting.

**Anticyclonic** High atmospheric pressure.

**Venturi effect** When air flows in a gap between two buildings, causing the wind to pick up high speeds.

■ hence, gusts of winds can often be much stronger in central urban areas

A single building can modify an airflow passing over it:
■ air is displaced upwards and around the sides of the building and is also pushed downwards in the lee of the structure
■ on the windward side air pushes against the building, creating relatively high pressure — this increases with height and causes a descending flow that forms a vortex when it reaches the ground and sweeps around the windward corners
■ air flowing around the sides of the building can become separated from the walls and roof and create suction in these areas

A group of buildings:
■ creates a disturbance to the airflow that depends on the height of the buildings and the spacing between them
■ if they are widely spaced, each building will act as an isolated block (as above)
■ if they are closer, the wake of each building interferes with the airflow around the next structure and produces a complex pattern of airflow that is difficult to predict

## Air quality

Air quality is a direct reflection of atmospheric pollution in urban areas:

■ the amount of air pollution depends on the rate at which pollutants are produced and the rate at which they are dispersed (diluted) as they move away from their source
■ the key atmospheric pollutants that are likely to have an impact on health are: ozone ($O_3$), nitrogen dioxide ($NO_2$), sulfur dioxide ($SO_2$) and particulate matter
■ a significant local source of air pollution is traffic emissions
■ photochemical smog is another problem — it also causes health problems (headaches, eye irritation, coughs and chest pains)

## Pollution reduction policies

Key policy measures:
■ atmosphere measures: Clean Air Acts, smoke-free zones and regulations on levels of airborne pollution, particularly on the level of $PM_{10}$ **particulates** in the atmosphere
■ vehicle control: traffic-free zones, pedestrianised areas, congestion charges, and restrictions by registration plate numbers — odd numbers one day, even numbers the next
■ public transport: persuading more people to use public transport rather than bring their cars into the city. Such schemes have

> **Exam tip**
>
> You may be able to illustrate these air movements by using diagrams — several textbooks use them.

> **Key term**
>
> Particulates Microscopic matter referred to as $PM_{10}$ (from exhausts, cement dust, tobacco smoke and ash) and $PM_{2.5}$ (fine particulate matter).

included the development of tram systems, bus-only lanes and park-and-ride schemes

- zoning of industry: placing industry downwind in a city; planning legislation has forced companies to build higher factory chimneys that emit pollutants above the dome layer
- vehicle emissions legislation: encouraging manufacturers to develop better fuel-burning engines and introduce catalytic converters to remove most of the particulates from exhaust fumes. Hybrid and electric cars should also have an impact

> **Exam tip**
>
> Be able to refer to named examples for some of these strategies.

> **Do you know?**
>
> 1 The UHIE tends to occur when there is a temperature inversion. What is a temperature inversion?
>
> 2 Outline the pollution problems that exist in named cities around the world.
>
> 3 Outline the vehicle restrictions that London has put in place to address air pollution.

# 6.4 Urban drainage

> **You need to know**
>
> - the impact of urban areas on hydrology
> - how sustainable urban drainage systems (SuDS) operate
> - a case study of a river restoration/conservation project

## Urban hydrology

Urban areas affect hydrology in a number of ways:

- impermeable roofs and roads are shaped to get rid of water quickly
- combined with a dense network of drains and sewers, this means that water gets to a river very quickly, reducing lag time and increasing discharge
- hydrographs in urban areas therefore are more flashy (steeper rising and recessional limbs) and have higher peak discharges

Flooding can be a serious problem:

- local authorities have to restrict the water passing through an area that has arisen from upland areas upstream
- surface water flooding is common because of larger proportions of impervious surfaces (concrete, tarmac, paved driveways)

> **Exam tip**
>
> Note that urban areas only exacerbate the impact on hydrology of the prevailing weather conditions over an area.

- intense rainfall on such surfaces readily ponds up before it can escape via man-made drainage systems
- these are sometimes inadequate, or become blocked by vegetation and/or litter
- to add to the risk, where man-made drainage systems are effective, surface water is channelled directly into drains and sewers, so the river rises quickly
- in times of spate, debris can build up directly behind bridge supports and exaggerate the effects of a flood

# Sustainable urban drainage systems

SuDS involve:
- constructed facilities and materials to store and drain water naturally:
  - permeable surfaces such as porous linings around trees and grassed areas
  - infiltration trenches, ponds and swales
  - underground storage
  - green roofs and walls
  - wetlands
- practices that involve management of water quality:
  - mitigation of pollution accidents
  - reduction of polluting activities

# River restoration/conservation

You are required to study a specific project of river restoration and/or conservation in a damaged urban catchment. This should include:
- the reasons for the project, including its overall aims
- contributions from the various parties involved, including their attitudes to the project
- the various ways in which restoration and/or conservation have been achieved
- an evaluation of the success or otherwise of the project

## Key term

**Sustainable urban drainage systems (SuDS)** An approach to managing rainfall in urban areas that seeks to replicate natural drainage systems, managing the rainfall close to where it falls.

## Exam tip

Be able to name some examples of SuDS schemes — some will be local.

## Exam tip

The choice of project is crucial here, and the parameters — that it is both urban and previously damaged (perhaps by industrial activity) — are vital. Examples may arise from any previously industrialised area such as the east of London or Salford Quays, Manchester.

## Do you know?

1 Identify three differences between an urban storm hydrograph and a rural one.
2 Describe the main features of one SuDS scheme you have studied.

# 6.5 Urban waste and other environmental issues

## You need to know

- the main sources of waste in urban areas and their relationship with the characteristics of people generating the waste
- the environmental impacts of approaches to waste management
- how incineration and landfill compare within one urban area
- the features of other environmental problems in urban areas, and how they can be addressed

## Sources of waste

Key points:

- the average person in the UK produces over 500 kg of household waste per year
- waste is also generated in huge amounts by industrial and commercial activity
- medical institutions create a specific type of waste, which needs careful processing
- a range of industries exist to manage such waste, however, some of which are international
- for example, several plants exist in the UK that recycle parts from cars and motor vehicles — they are much more sophisticated than 'scrap yards'
- some trade in industrial waste is difficult to follow — an example is the shipment of electronic waste around the world
- many people feel that the amount of waste generated is unsustainable and that we need to change our attitude to it — to see it as a resource to be managed rather than as a nuisance

### Exam tip

The movement of waste around the world mirrors that of other aspects of the global system — from the developed to the developing world.

## Waste disposal

See Table 20.

Table 20 Methods of waste disposal

| Method | Commentary |
|---|---|
| Unregulated | Waste is placed in illegal locations — any disposal site — which could include hot/dirty water into rivers; landfill is sometimes said to be unregulated |
| Recycling | The reprocessing of waste — metals, plastics, paper — into new products is now a major industrial activity in many countries, both formally and informally; targets — by the EU, the UK and local authorities — for recycling are now common practice<br><br>Reducing the amount of packaging and plastic bags is a form of recycling |
| Recovery | The reuse of organic materials, using digestive energy-producing plants or composting of organic waste; needs careful management as the by-products can be dangerous |
| Incineration | A form of energy recovery by burning waste; most sites in the UK use the heat created to warm buildings or generate electricity; there are some serious concerns regarding atmospheric pollution, denied by companies |
| Burial (landfill) | The waste is buried — dumped in old quarries or hollows where it is unsightly and a threat to groundwater supplies and river quality as toxic chemicals are leached out; decomposing waste also emits methane, the most toxic of the GHGs and potentially explosive; to discourage use of landfill sites, landfill taxes are often imposed, and the activity is now closely regulated in the UK |
| Submergence | The 'burial' of waste at sea is banned by international convention; illegal dumping of ship oil still takes place |
| Trade | Electronic waste is shipped around the world, from the developed to the developing world; this trade in industrial waste is difficult to follow |

**Exam tip**

You are required to carry out a comparative study of incineration and landfill in a specified urban area. There are few cities in the UK where both systems operate. Sheffield and Douglas (Isle of Man) are possible areas of study. Overseas, you could investigate Amsterdam or Singapore.

# Other environmental issues

## Atmospheric pollution

See Air quality (page 112).

## Water pollution

Sources:

- runoff from streets carries oil, rubber, heavy metals and other contaminants from cars and other vehicles
- untreated/poorly treated sewage water can be low in dissolved oxygen and high in pollutants such as faecal coliform bacteria, nitrates and phosphorus

- treated sewage can still be high in nitrates
- warm water from power stations — thermal pollution
- groundwater and surface water can be contaminated from waste dumps, toxic waste and chemical storage areas, leaking fuel storage tanks, and intentional dumping of hazardous substances
- domestic water drains can carry cooking oil, paints, detergents and litter (e.g. baby-wipes)
- misconnections — misconnected properties let **greywater** enter surface water drains
- one-off incidents such as a road tanker accident

**Key term**

**Greywater** Polluted water originating from bathroom sinks, showers, bathtubs and washing machines.

Consequences:
- oils, cooking oils and fats spread out across the top of a watercourse and cause a rainbow effect called iridescence
- larger amounts can create a matt effect on a water surface, and pools of oil may weather and solidify
- greywater can cause algae and fungi to grow at outfalls and in watercourses
- polluting substances deplete oxygen from water, causing living things to 'suffocate' — they can also be directly toxic to animals and plants

Strategies:
- solutions involve sustainable ways for an urban area to:
  - □ reduce its dependence on pollutants and the amount of pollutants it produces
  - □ recycle or dispose of pollutants before they contaminate water
  - □ educate industries and individuals in ways to reduce polluting activities
- specific strategies include:
  - □ treatment plants
  - □ legislation and enforcement
  - □ combining with SuDS to improve water quality naturally

# Dereliction

Problems with **dereliction**:
- negative perspective created of an area — vandalism increases, house prices nearby fall
- land is often contaminated and dangerous to trespassers
- areas are colonised by plants that can be challenging to remove — e.g. Japanese knotweed spreads easily via rhizomes and cut stems or crowns; it is now listed under the Wildlife and Countryside Act 1981 as a plant that is not to be planted or otherwise introduced into the wild

**Key term**

**Dereliction** Land that has been abandoned, such as after the clearance of former industrial sites, and then becomes dilapidated.

Strategies:

■ clearance as part of urban regeneration schemes

■ new building developments on 'brownfield' sites

■ remediation — removal of pollutants and toxins

■ community action — e.g. small-scale urban farms

## Do you know?

1 Describe the main features of the movement of electronic waste around the world.

2 Outline one small-scale water-quality management scheme.

3 Outline the remediation of one former derelict area.

# 6.6 Sustainable urban development

## You need to know

■ how urban areas impact on local and global environments, with particular reference to 'ecological footprint'

■ the nature and features of sustainable cities, including the concept of liveability

■ strategies for developing more sustainable cities

## Key terms

**Sustainability** Meeting the needs of today without compromising the ability of future generations to meet their own needs.

**Ecological footprint** A measurement of the area of land or water required to provide a person (or society) with the energy, food and other resources they consume and to render the waste they produce harmless.

## Sustainability

Cities pose a threat to both the local and the global environment:

■ by creating pollution (air, water) and dereliction they threaten **sustainability**

■ they can have a very high **ecological footprint**

Consequently, there have been pressures to enshrine sustainability in cities (Table 21).

Table 21 Dimensions of sustainability

| Economic | Social | Environmental | Governance |
|---|---|---|---|
| Individuals and communities should have access to a reliable income over time<br><br>Employment opportunities for all | All individuals should enjoy a reasonable quality of life, with access to education, food, health services, clean water and sanitation | No lasting damage should be done to the environment<br><br>Renewable resources must be managed in ways that guarantee continued use | Commitment to sustainable policies by decision makers, such as 'green' planning policies<br><br>Strategies to reduce inequalities |

Linked to the above is the concept of **liveability**:

■ a combination of economic, aesthetic and environmental factors provides a much more balanced perspective on a city's 'value' to its citizens

■ an index of liveability can be created using measures of economic vibrancy and competitiveness, domestic security and stability, sociocultural conditions, public governance, environmental friendliness and sustainability

■ in 2016, the five best cities to 'live in' were judged to be Melbourne, Vienna, Vancouver, Toronto and Calgary

### Key term

**Liveability** A measure of what it is like to live in a city and how urban life there compares with other cities in the world. It assesses 'work–life balance', environmental awareness, and a sense of localism vs globalism.

# Sustainable cities

Features:

■ see Figure 44

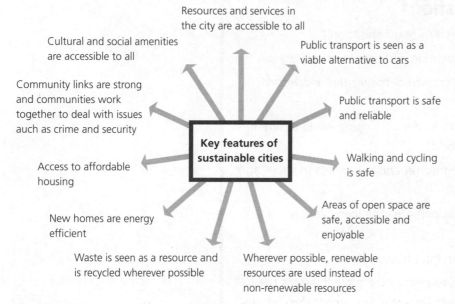

**Figure 44 Features of sustainable cities**

Strategies:

■ increasing the use of renewable sources of electricity

■ appropriate systems of waste management, including dealing with food waste

■ reducing fuel emissions within urban areas — buses, lorries, railways, airports

■ investment in affordable and low-carbon housing

■ establishing household goods recycling — investment in reuse schemes and provision of space for reuse activities

■ reduced consumption of water — wastage through pipe leaks can be a huge problem

### Exam tip

You are required to study case studies of two contrasting urban areas to illustrate and analyse all of the key themes set out here, to include:

■ patterns of economic and social wellbeing

■ the nature and impact of physical environmental conditions

This must be done with particular reference to the implications for environmental sustainability, the character of the study areas, and the experience and attitudes of their populations.

Possible examples include: London, Liverpool, Johannesburg, Mumbai and Barcelona.

- greater use of public transport — e.g. the rollout of technologies such as hydrogen fuel cell buses and hybrid electric buses, electrification of the railway network and biodiesel for trains
- local sourcing — choosing the most local supplier for standard products
- 'green governance' — ensuring all decision-making is aimed at sustainable outcomes

## Do you know?

1 When and where did the concept of sustainability first appear?

2 Name some examples of sustainable cities.

3 It has been suggested that adopting a local currency can make an urban area more sustainable. Explain this.

## End of section 6 questions

1 Describe the current global pattern of urbanisation.

2 Distinguish between suburbanisation and counterurbanisation.

3 Outline and comment on the growth of megacities and world cities.

4 In what ways is urban structure in the developing world different from that in the developed world?

5 Assess the extent to which counterurbanisation leads to social and economic change.

6 Assess the role of perception in perpetuating inequality in developing world cities.

7 Describe and explain the urban heat island effect.

8 With reference to a specific river-restoration project, assess the extent to which it has an impact on water movement through the urban catchment.

9 What can be done to combat atmospheric pollution in urban areas?

10 To what extent can urban areas be sustainable?